151
Advances in Biochemical Engineering/Biotechnology

Aims and Scope

This book series reviews current trends in modern biotechnology and biochemical engineering. Its aim is to cover all aspects of these interdisciplinary disciplines, where knowledge, methods and expertise are required from chemistry, biochemistry, microbiology, molecular biology, chemical engineering and computer science.

Volumes are organized topically and provide a comprehensive discussion of developments in the field over the past 3–5 years. The series also discusses new discoveries and applications. Special volumes are dedicated to selected topics which focus on new biotechnological products and new processes for their synthesis and purification.

In general, volumes are edited by well-known guest editors. The series editor and publisher will, however, always be pleased to receive suggestions and supplementary information. Manuscripts are accepted in English.

In references, Advances in Biochemical Engineering/Biotechnology is abbreviated as *Adv. Biochem. Engin./Biotechnol.* and cited as a journal.

More information about this series at http://www.springer.com/series/10

Georg M. Guebitz · Alexander Bauer
Guenther Bochmann · Andreas Gronauer
Stefan Weiss
Editors

Biogas Science and Technology

With contributions by

T.M. Callaghan · C. Carliell-Marquet · G. Collins
V. Dollhofer · C.D. Dubé · G. Esposito · F.G. Fermoso
K. Fliegerová · B. Fröschle · L. Frunzo · T. Gehring
G. Wyn Griffith · G.M. Guebitz · G. Guibaud · S.R. Guiot
M. Heiermann · K. Koch · P. Kosse · M. Lebuhn
M. Lübken · U. Messelhäusser · B. Munk · M. Plöchl
S.M. Podmirseg · S.K.M.R. Rittmann · B.H. Svensson
E.D. van Hullebusch · J.P.M. Vink · S. Weiß
M. Wichern

Springer

Editors
Georg M. Guebitz
Institute of Environmental Biotechnology
BOKU Vienna—University of Natural Resources and Life Sciences
Tulln
Austria

Alexander Bauer
Institute of Agricultural Engineering
BOKU Vienna—University of Natural Resources and Life Sciences
Tulln
Austria

Guenther Bochmann
Institute of Environmental Biotechnology
BOKU Vienna—University of Natural Resources and Life Sciences
Tulln
Austria

Andreas Gronauer
Institute of Agricultural Engineering
BOKU Vienna—University of Natural Resources and Life Sciences
Tulln
Austria

Stefan Weiss
Institute of Environmental Biotechnology
BOKU Vienna—University of Natural Resources and Life Sciences
Tulln
Austria

ISSN 0724-6145 ISSN 1616-8542 (electronic)
Advances in Biochemical Engineering/Biotechnology
ISBN 978-3-319-35880-2 ISBN 978-3-319-21993-6 (eBook)
DOI 10.1007/978-3-319-21993-6

Springer Cham Heidelberg New York Dordrecht London

Softcover reprint of the hardcover 1st edition 2015

Printed on acid-free paper

Springer International Publishing AG Switzerland is part of Springer Science+Business Media (www.springer.com)

Contents

Microbiology and Molecular Biology Tools for Biogas Process Analysis, Diagnosis and Control

Michael Lebuhn, Stefan Weiß, Bernhard Munk and Georg M. Guebitz

Abstract Many biotechnological processes such as biogas production or defined biotransformations are carried out by microorganisms or tightly cooperating microbial communities. Process breakdown is the maximum credible accident for the operator. Any time savings that can be provided by suitable early-warning systems and allow for specific countermeasures are of great value. Process disturbance, frequently due to nutritional shortcomings, malfunction or operational deficits, is evidenced conventionally by process chemistry parameters. However, knowledge on systems microbiology and its function has essentially increased in the last two decades, and molecular biology tools, most of which are directed against nucleic acids, have been developed to analyze and diagnose the process. Some of these systems have been shown to indicate changes of the process status considerably earlier than the conventionally applied process chemistry parameters. This is reasonable because the triggering catalyst is determined, activity changes of the microbes that perform the reaction. These molecular biology tools have thus the potential to add to and improve the established process diagnosis system. This chapter is dealing with the actual state of the art of biogas process analysis in practice, and introduces molecular biology tools that have been shown to be of particular value in complementing the current systems of process monitoring and diagnosis, with emphasis on nucleic acid targeted molecular biology systems.

Keywords Biogas process parameters · Molecular biology tools · Quantitative Real-Time PCR · Next generation sequencing · Meta-omics · Fluorescence-in situ hybridization · Metabolic quotient

---

M. Lebuhn (✉) · B. Munk
Department for Quality Assurance and Analytics, Bavarian State Research Center for Agriculture (LfL), Lange Point 6, 85354 Freising, Germany

S. Weiß
Austrian Centre of Industrial Biotechnology (ACIB), Petersgasse 14, 8010 Graz, Austria

G.M. Guebitz
Institute of Environmental Biotechnology, BOKU Vienna—University of Natural Resources and Life Sciences, Konrad Lorenz Str. 20, A-3430 Tulln, Austria
e-mail: guebitz@boku.ac.at

G.M. Guebitz et al. (eds.), *Biogas Science and Technology*,
Advances in Biochemical Engineering/Biotechnology 151,
DOI 10.1007/978-3-319-21993-6_1

Abbreviations

BB	Bead-beating
BMP	Biological/biochemical methane potential
BLAST	Basic local alignment search tool
Bp	Base pair(s)
cDNA	Complementary DNA (transcribed from RNA species)
CLSM	Confocal laser scanning microscopy
COD	Chemical oxygen demand
DGGE	Denaturing-gradient gel electrophoresis
DNA	Deoxyribonucleic acid
FISH	Fluorescence in situ hybridization
LCB	Lignocellulosic biomass
LM	Light microscopy
MQ	Metabolic quotient
mRNA	Messenger RNA
NA	Nucleic acid(s)
NGS	Next generation sequencing
OLR	Organic loading rate
PC(o)A	Principal coordinate/Principal component analysis
PCR	Polymerase chain reaction
PSM	Process simulation model
qPCR	Quantitative Real-Time PCR
rDNA	Ribosomal deoxyribonucleic acid
RNA	Ribonucleic acid
rRNA	Ribosomal ribonucleic acid
RT	Reverse transcription
SCFA	Short-chain fatty acid(s) or also VFA
SEM	Scanning electron microscopy
SMA	Specific methanogenic activity
TEM	Transmission electron microscopy
TGGE	Temperature-gradient gel electrophoresis
TVA/TIC	Total volatile acids/total inorganic carbon
VFA	Volatile fatty acids
VOA	Volatile organic acids

Contents

1 Introduction

Biogas production by anaerobic digestion of organic matter is a bio-technology with very long tradition for some 2,000–3,000 years. It was applied initially for sanitation purposes and only later additionally for energy production. The issue sanitation with its beneficial effects for the society is presented within this book in Chap. 3. All of the process steps are performed in a food chain by different microorganisms, governed by process engineering in a suitable technical environment. Some of these microbes have to cooperate extremely efficiently in syntrophic dependency in order to be able to thrive and proliferate at the minimum limit of possible energy gain [1, 2].

Methanogenic archaea, and among these particularly the acetoclastic *Methanosaetaceae*, appear to be most sensitive in biogas processes to stress factors such as short retention times, high ammonia, oxygen and short-chain fatty acid (SCFA) concentration, lack of certain trace elements and increased temperature [3, 4]. Due to their relatively low apparent maximum turnover number (K_m) for acetate and long doubling times [5], the acetoclastic methanogens are disfavored at short retention times and increasing acetate concentration in the fermenter [6]. They are increasingly washed out if their proliferation cannot compensate out-dilution. This effect is even pronounced at additional stress conditions, favoring the activity and growth of syntrophic associations with hydrogenotrophic methanogens to the detriment of active *Methanosaetaceae* and acetoclastic activity [3, 4, 7]. It is incorporated as a central point in the bioindicator concept of process diagnosis [4] (see also Sects. 3 and “Microbial Guilds, Bioindicators and Transcriptional Profiling”).

The second bottleneck is the thermodynamically difficult hydrogen, formate or electron-releasing conversion of short chain fatty acids (SCFAs), alcohols and other intermediates of the biogas process. Most of these reactions are endergonic at standard conditions but can be realized by syntrophic associations involving product-scavenging methanogens [8, 9]. Methanogenic archaea are able to remove the reaction products by converting them finally to biogas, predominantly CH_4 and CO_2, which segregates from the fermenter sludge to the gas headspace and is further withdrawn by gas utilization. Syntrophic bacteria or anaerobic fungi partners of methanogens are difficult to cultivate and to study without their product-consuming associate. Modern characterization is typically initiated by genome or metagenome analysis, possibly leading to insights about special requirements that allow cultivation of pure isolates and studying their special physiological performances [10–12].

A third recognized bottleneck is the initial rate-limiting hydrolysis of recalcitrant substrates such as lignocellulose-rich biomass (LCB). When compared to aerobic degradation of lignocellulose, considerably less is known on the corresponding anaerobic process and the organisms involved. Besides bacteria, other organisms such as anaerobic fungi may be involved in efficient initial LCB attack and degradation [13]. Chapter 2 in this book is dedicated to anaerobic fungi and recent perceptions of their role in anaerobic LCB digestion. For some of these cellulolytic organisms, the genome has been sequenced [14, 15]. Such genome information is an invaluable data basis for process optimization and further biotechnological exploitation.

Microbial processes in anaerobic digestion are driven by both, biotic and abiotic factors. The physical and chemical environment (e.g. nutritional factors and redox status) are basic to and determine the biotic activity, the substrate conversion by the microbes. Biotic measures to regulate the process (e.g. bioaugmentation) however are scarce as briefly discussed in Sect. 2.7.

The most important issue in process optimization is to avoid the worst case, process disturbance or even breakdown. This requires a process control strategy that includes reliable process diagnosis based on meaningful analytical data. Since the activity of bioindicator microbes, organisms that are typical for certain process conditions, does react before conventionally used process chemical parameters indicate process failure, a promising approach for successful process control is to assess the activity of these bioindicators as integral part of an early-warning system [4]. The relevant actors, i.e. bioindicators performing the crucial biogas process steps, must hence be identified, and suitable analysis tools must be used or developed to track these key organisms and their activity quantitatively.

In the following chapters, microbiology and molecular biology tools for biogas process diagnosis and control are compiled and discussed. Since several important process dynamics such as SCFA and total solid (TS) turnover as well as gas quality/quantity are the result of microbial activity, and respective wet chemistry and physico-chemical analyses are and will be indispensable part of conventional practice but have revealed limitations, recent experience with these conventional applications for agricultural single-stage biogas processes is presented in the following Sect. (2). Molecular biology approaches have only recently emerged and may be introduced into practice after comparison or along with established physico-chemical routines. Some of these molecular tools, however, are promising candidates to be implemented in a holistic suite of analytical tools for process diagnosis and control.

2 Physico-Chemical and Biochemical Process Parameters

The spectrum of physico-chemical parameters actually employed for process diagnosis of agricultural and category 2 biowaste (untreated non-infectious to humans, animals or plants), biogas plants has originally been adopted from anaerobic sewage sludge digestion. Many of these parameters and respective benchmarks are listed and discussed in a review by Weiland [16]. Important aspects for diagnosis and control of single-stage processes are presented in the following.

2.1 Gas Production

In order to evaluate the efficiency of the process it is indispensable to determine the volumetric gas production and the gas quality or at least estimate these parameters from the generated electricity and the actual adjustment of the combined heat and power unit. Different equipment is on the market, ranging from simple manually operated lab instruments to fully automated industry scale online devices. Combined with data on the fed organic dry matter (oDM or volatile solids [VS]), the gas production and quality data inform on the specific methane production or methane yield (m^3 CH_4/kg VS). Comparison with benchmarks for given substrates and interpretation of the recent methane yield development allows to estimate the actual process efficiency at least roughly and to reveal up- and downward trends. Particular attention should be paid to decreasing CH_4 and increasing H_2 concentrations in the produced biogas. CH_4 concentrations falling below ca. 48 % and H_2 concentrations exceeding ca. 100 ppm in single stage processes are alarming and should give rise to counteractive measures.

Isotope ratio mass spectrometry (IRMS) based methods analyzing isotope discrimination by the biogas producing microbial community were recently proposed to detect methanogenic pathway shifts [17, 18]. The switch from acetoclastic to hydrogenotrophic methanogenesis is interpreted as signal of stress conditions which may allow plant operators to adjust their feeding strategy. Since the IRMS equipment is expensive, a laser-assisted online analysis technique was described for this purpose as a more practice-oriented alternative [19]. However, a pathway shift does not necessarily indicate imminent process failure, and interpretation problems with data from variable feedstock composed of C4 and C3 plant material still need to be resolved. Although gas analysis using stable isotope ratios has potential to reveal biogas production pathway changes, and online monitoring is possible, its contribution to process diagnosis is thus confined to basic research in its current state.

2.2 Process Intermediates, SCFA, Total and Volatile Solids, and Specific Determinants

The determination of SCFA (also referred to as volatile fatty acids [VFA], or volatile organic acids [VOA]) is a highly important component of process diagnosis. The SCFA spectrum is typically assessed using liquid or gas chromatography (LC/GC) based routines on suitable extracts in an external specialized lab and should include the iso-forms of butyrate and valerate. Increased levels of these

SCFA exceeding ca. 50 mg/L as well as propionic acid concentrations above ca. 1 g/L along with a propionic/acetic acid ratio >1 typically indicate process disturbance in single-stage systems [16], but exceptions have become known as well. Reduced activity of methanogens due to substrate overload or limited availability of essential nutrients such as trace elements is frequently the reason, giving rise to "acid jam", i.e. accumulation of upstream produced intermediates.

Attention must also be paid to the development of the dry matter (or total solids [TS]) content in the digester. TS can easily be determined on-site in an oven at 105 °C, whereas for analysis of VS, a muffle furnace is required. Information on VS contents in the substrates is important for determining the methane yield, and the VS/TS trend in the digester can anticipate eventually problematic ash accumulation. Increasing TS values over time in the fermenter sludge indicate a problem at the hydrolysis/acidogenesis step leading to compromised process efficiency and incomplete digestion. TS values exceeding 15 % can lead to stirring problems in conventional continuously stirred tank reactors (CSTRs). Reducing the organic loading rate (OLR), i.e. increasing the hydraulic retention time (HRT), might be helpful, otherwise substrate conditioning by physical/mechanical or (bio)-chemical means could be considered. More specific information of TS and VS in the fermenter sludge can be obtained by the fractionated analyses according to Weende and Van Soest [20].

Recently, an online near-infrared spectroscopy (NIRS) application to evaluate the process state with potential for the practice was reported [21]. The authors presented acceptable estimations of VS, ammonium, total inorganic carbon (TIC) and total VFA even in short-term process dynamics of a mesophilic pilot-scale maize silage fed biogas digester. Further extension of the model to include different substrates and analysis parameters, and experience in long-term operation is needed before online NIRS systems can be recommended for process control.

Similarly, techniques involving flow-assisted cell sorting (FACS) [22], matrix-assisted laser desorption/ionization—time of flight mass spectrometry (MALDI-TOF/MS) and Fourier transform ion cyclotron resonance mass spectrometry (FT-ICR/MS) [23, 24] as well as secondary-ion mass spectrometry (SIMS) based systems [25, 26] can be helpful to separate or identify distinct microbes or consortia, or track their specific metabolic activities. Such methods prove to be useful in basic microbiology research and taxonomy and allow identifying microbes with resolution at the subspecies or strain level [23], given a suitable reference database is available. However, "dirty" environmental samples can pose considerable problems. Although there is some potential, application for monitoring the process status of the black-box biogas fermenter actually does not appear to be a realistic option. Physical methodologies basing on or coupled with fluorescence-in situ-hybridization (FISH) are itemized in Sect. 3.2.1.

2.3 Early Warning—The TVA/TIC Ratio

The ratio of total volatile acids to total inorganic carbon (TVA/TIC, also referred to as FOS/TAC or VOA/TIC) is determined by 2-point titration (pH 5.0, pH 4.4) and can easily be performed on-site [27]. It beats out the pH value as early indicator of process acidification due to its much higher sensitivity. Currently, the TVA/TIC ratio is the most used process chemical early warning system of acidification. TVA/TIC ratios of 0.15–0.45 are typical for a stable process without major acid accumulation, whereas rising ratios exceeding 0.45 reflect process disturbance, and values above 0.6–0.7 indicate acidosis. This can be associated with TIC depletion, e.g. in case of trace element deficiency.

However, in cases of atypical process conditions such as at higher NH_4^+ concentrations and pH-values, considering only TVA/TIC as a process indicator can be risky. The NH_4^+/NH_3 buffer system can trap protons masking possible acidification events. Obtaining low TVA/TIC values can thus be misleading at higher free ammonia-nitrogen (FAN; see also Sect. 2.4) if distinct SCFA (possibly not measured) may already be at alarming level. Above ca. 1 g NH_4^+-N/L, it is therefore suggested to monitor the SCFA spectrum and/or molecular biology parameters (see Sect. 3) as well in order to perceive process perturbation and acid build-up.

2.4 Nutrients, Toxic and Disturbing Agents

Since nutrient composition of the substrates governs and limits microbial process performance, respective analysis should be performed occasionally and particularly if a plant is operated with atypical substrates or such operation is planned. Modern elementary analysis involves Inductively Coupled Plasma (ICP) equipment with detection by Optical Emission Spectroscopy (OES) or even more sensitive Mass Spectrometry (MS). Experience from practice suggests that the C:N ratio should be about 15–45 and the C:N:P:S ratio about 300-600:15:5:3 [4, 16, 28].

Several toxic agents are known that can impede the anaerobic digestion process. One of the most relevant is free ammonia [29] (see also Sect. 2.3) which can be calculated from the NH_4^+-N concentration, the process temperature and the pH value [30]. Nitrogen seems to be lacking only in exceptional cases, but reduced N-compounds typically accumulate in anaerobic digestion of protein-rich feedstock and can become toxic [6]. It is thus important to determine these parameters periodically. NH_3 diffuses unspecifically inside the cell, can capture protons and hamper proton-dependent ATP generation leading to activity loss and possibly cell death. The typical ammonia toxicity threshold is about 400–500 mg NH_3-N/L, but a higher margin is possible in case of adaptation [4, 29]. Particularly microbes relying on H^+ pumps are susceptible whereas those with Na^+-pumps are favored in the presence of sufficient Na^+. Predominantly microbes involved in the hydrogenotrophic metabolism of biogas intermediates are using Na^+ pumps for ATP generation [5, 9]. Na^+ appears to limit methanogenesis and elicit acidification at values decreasing to about 10 mg/L in the fermenter sludge and must therefore be provided at higher concentration in the substrate mix [31]. This can explain stabilizing, stimulating effects of Na^+ addition at limiting, constrained or stress conditions such as in high-performance biogas production from grass silage in the practice [4].

Requirement for trace elements (TEs) in biogas production from biomass, particularly of Co, Ni, Se, Fe, and possibly of Mo, B and W, has been described in many publications, e.g. [16, 32–34] and in Chap. 7 of this book. Since their presence in suitable, available concentrations is a precondition of efficient process performance, TE contents should be determined occasionally and particularly if the feedstock composition is changed. Trace elements, however, can become toxic in higher concentration. Other compounds with toxic or disturbing potential are found among e.g. antibiotics, mycotoxins, detergents and heavy metals [35, 36], and some phenolic compounds appear to have inhibitory properties [37]. Cu and Zn loads are of particular importance for agricultural biogas plants. They can originate in higher concentrations from animal husbandry, and according to several practice reports, can be the cause of process disturbance and efficiency loss [38].

Several devices allow the measurement of O_2 and H_2S in the biogas. O_2 can enter the process by leakages or actively during biological desulfurization. CO_2 reduction becomes unfavorable in the presence of better electron scavengers such as O_2, SO_4^{2-} and NO_3^- [6]. The redox potential is increased in their presence, and the activity of most anaerobic microbes is impeded. O_2 should therefore be kept below 1 % (better 0.1 %) in the gas phase, and feeding substrates with high SO_4^{2-} and NO_3^- contents minimized, for similar reasons. H_2S typically originates from sulfur containing organic matter. Since it is highly toxic for most living beings, maximum working place concentrations have to be respected. Moreover, corrosive acids such as H_2SO_4, H_2SO_3 or HNO_3 can be formed in the presence of O_2 and S- or N-containing compounds. They can damage mechanical devices and constituents of the biogas plant.

2.5 *Biological Methane Potential, and Activity, Toxicity and Supplementation Tests*

BMP (biological methane potential) or SMA (specific methanogenic activity) tests are typically applied to determine the methane potential of given substrates. By variation of these batch-mode assays in ATS (activity, toxicity and supplementation) tests [39], they can be employed to diagnose and control the actual state of the biogas process. It is attempted to assess e.g. the capacity of inocula to be activated, to evaluate the potential of added or endogenous compounds to exert toxic effects, or to test supplements for process stimulation.

Depending on the type of inoculum and its degree of adaptation, the potential of the biocenosis is tested to produce methane and/or react to changed process conditions. It is emphasized that results of these assays cannot always be used to predict the performance such as the methane yield in flow-through operation. This is mainly due to operational differences, particularly in the effective organic loading rate and the actual microbial retention time. Moreover, these tests are labor- and time consuming, typical test periods vary between several days to weeks.

2.6 Enzyme Tests and Applications

Enzymatic tests are important in research and have some potential for practice application. For example, hydrolytic enzymes are of high interest, since enzymatic saccharification (hydrolysis) is a rate-limiting step in anaerobic digestion (AD) from solid substrates and especially undigested lignocellulosic biomass (LCB) material, e.g. floating layers can pose considerable operational problems [40–42]. Hydrolases must accommodate heterogeneous plant cell wall residing polymers with various degrees of polymerization (DPs), side chain branching patterns and several altering substitutes [43]. Because of this chemical inhomogeneity and substrate specificity corresponding enzymes that act upon them are generally difficult to isolate and characterize. The quantitative determination of enzymatic activities is commonly based on accumulated products after hydrolysis including reducing sugars, total sugars and chromophores. Other assays measure the reduction in substrate quantity or the change in the physical properties of substrates. However, the production of reducing sugars is assayed using alkaline dinitrosalicylic acid (DNS), copper-arsenomolybdate using the 4-hydroxy-benzoylhydrazine (PAHBAH) method, 2,2′-bicinhroninate (BCA) and ferricyanide or directly using anthrone- or phenol-H_2SO_4. Focusing monomeric products, i.e. glucose as major product, commercial enzymatic glucose kits using coupled hexokinase and glucose-6-phosphat dehydrogenase are available. The main drawback of these methods is a poor stoichiometric relationship between reaction products (e.g. cellodextrins, malto- or xylodextrin) and pure D-glucose standards [44], which may result in an underestimation or overestimation of cellulase and hemicellulase activities [45, 46]. However, substrates used for hydrolysis assays should therefore always be as similar to native polymer structures as possible in matters of DP, solubility and crystallinity.

The esterase activity was suggested as indicator for the overall fermentation process, representing a sum parameter for bacterial heterotrophic activity in general [47]. In this context a positive correlation between esterase activity and substrate conversion rate towards methane was observed, revealing that process disruption is reflected by decreased enzyme activities [48, 49]. Furthermore, a negative correlation of esterase as well as aminopeptidase activities and substrate quality was observed, providing fermentability indications regarding silage as substrates [48]. Therefore, enzyme assays can be a useful tool for monitoring the overall anaerobic digestion process. Modifications towards an all-in-one testing kit like available for other chemical parameters (e.g. Merck Spectroquant® for COD, TOC, nitrate, ammonium etc.) is desirable to provide plant operators with an activity specific easy-to-use monitoring instrument. Compared to photometric tests, lower detection limits (factor 20–500) and shorter reaction times can be reached by the use of fluorimetric determination on the basis of fluorigenic compounds (e.g. fluoresceine diacetate, azocasein) [50]. Thereby, a precise study of catabolic enzyme activities such as esterase, phosphatase, aminopeptidase and glucosidase activities in samples with low biomass density is possible. In sum, non-methanogenic (hydrolytic, acidogenic) and methanogenic activity tests in combination with molecular tools seem to be essential for a better characterization and monitoring of full-scale anaerobic digesters [51].

Beyond measuring the activity or amount of enzymes available in the process, enzymes can be also applied to stimulate the AD process. The utilization of enzymes for environmental and industrial applications have been described to be stable in a large range of even quickly changing conditions, i.e. pH, temperature, presence of inhibitors or interspecies competition [52, 53], although controversially observations have been described by other studies on the use of commercial enzymes and enzyme mixtures considering related costs in AD processes, rather suggesting a specific application with respect to optimum conditions and the source of substrate [54, 55]. It has been shown that a combination of chemical and enzymatic pretreatment of bamboo waste, using commercial cellulase and alkaline, can lead to significantly enhanced chemical oxygen demand (COD) solubilization and substrate saccharification in BMP tests, which not necessarily translates to high methane yields as compared to alkaline pretreatment alone [56], suggesting to re-think the role of enzymes in multiple-pretreatment settings. The application of natural endogenous hydrolases such as amylase and protease from fermentation sludge for pretreatment of wastewater sludge resulted in improved sludge solubilization and acidification regarding the COD and VFA upturn [57], whereas a positive effect on anaerobic biodegradability, hydrolysis, digestion rates as well as maintaining a healthy microbial population were not indicated [58]. However, enzyme treatment can improve the economic production of biogas from agricultural residues, municipal solid and animal wastes by enhancing the fluidity of fibrous feedstock mixtures [59], solubilization and deflocculation of wastewater and sewage sludge biomass towards anaerobic digestibility [60–62].

The instability and time-limited effect of free enzymes can be overcome by immobilization using suitable carriers such as alginate or minerals [52, 63–66]. Moreover, the improvement of high-solid substrate degradation can be also achieved by inoculation of beneficial bacteria, which produce corresponding hydrolases in response to the given feedstock and operating conditions [49, 67, 68]. This has been demonstrated for mixed hemicellulolytic bacteria cultures [69, 70] and isolated bacterial species obtained from natural biogas-producing consortia as well, i.e. hydrogen-producing cultures of *Caldicellulosiruptor saccharolyticus* and *Enterobacter cloacae* [71, 72] or *Clostridium cellulolyticum*, which was successfully adopted to enhance the hydrolysis of wheat straw leading to increased BMP tests improving the utilization of lignocellulosic substrates [73].

An improved understanding of the catalytic potential of the AD ecosystem can be attained by mechanistic models based on kinetic data capturing important details of enzyme-substrate interactions, key substrate surface properties and individual enzyme adsorption and complexation characteristics as demonstrated for cellulose/cellulase interactions [57, 74, 75]. These catalytic information might be implemented into existing complex dynamic models and simulations such as IWA's Anaerobic Digestion Model (ADM No. 1, 2) or novel Process Simulation Models (PSMs), which are validated against a variety of lab and industrial data on anaerobic digestion to predict the applicability of any substrate for biogas production at any given process condition [76–79].

2.7 *Classical Microbiology Approaches*

Classical cultivation-based microbiological methods have not gained major importance in the analysis of the biogas process status in practice. This is owed to the fact that most anaerobic microbes have long duplication times. Cultivation of anaerobes is not only tedious and difficult in many instances. It can cause biased results if specific growth and activity requirements of investigated microbes or associations are not known. By applying next generation sequencing (see Sect. 3.1.4), metagenomics, genome analysis and mapping, specific genetic capacities of investigated microbes or associations can be identified which can help to meet unrecognized cultivation requirements and eventually grow hitherto uncultured organisms [80]. Classical light microscopy reveals its limits given the highly turbid sample matrix and the low portion of known and described micro-organisms [81].

Bioaugmentation is a classical microbiology measure to counteract process imbalances and a key component of biotechnology routines. Virtually every biogas plant and biotechnological process has been or is started up by inoculation, a special form of bioaugmentation. Numerous experiences show that once a stable process and biocenosis is established, newly introduced strains will encounter enormous difficulties to colonize and propagate in this process [82], although such success was announced in a few reports [71]. Bioaugmentation might be helpful for the case that a disturbed process should be stabilized or re-established by the (re)introduction of certain strains or consortia which had been recognized to be relevant for proper function but were washed out.

3 Molecular Biology Approaches

Biogas process failure can have several reasons. Technical reasons include stirring problems, leakages and temperature changes. At too short microbial retention time and unbalanced or insufficient nutrition (see Sect. 2.4) slowly growing, but possibly important microbes are diluted out. If a process-relevant guild is washed out and no functional substitutes can grow up, this results in process failure or even breakdown. Such bioindicators of the process state are ideally tracked by specific molecular biomarkers. Since these react earlier than the conventional physico-chemical parameters [83], molecular biology bioindicator tracking does not only allow for diagnosing the process, it provides more time to plant operators for specific counteraction.

This chapter is subdivided into several sections where molecular biology methods with more or less potential for process diagnosis are described. Some are used only in basic research and others have started to be applied in practice.

3.1 PCR Based Approaches and Nucleic Acid Sequencing

Since the invention of the Polymerase Chain Reaction (PCR) in 1983, PCR based techniques have conquered the field in molecular biology diagnostics. With the recent progress in affordable next generation sequencing techniques [84] (see Sect. 3.1.4), sequence information in databanks has substantially boosted. On this basis, group-specific primers and probes can be designed with much higher dependability. Diagnostic PCR assays are quickly performed and highly sensitive if suitable (e.g. fluorescence based) detection systems are included. In this chapter, emphasis is therefore on PCR-based methods and among these particularly on quantitative Real-Time PCR (qPCR) assays. PCR applications typically need a specialized laboratory environment but developments for on-site use are emerging. On-line systems, however, are far from being conceivable.

3.1.1 Crucial Prerequisites: Sampling and Nucleic Acid Extraction

A prerequisite for reliable results is that the samples taken are representative of the fermenter sludge. This is not trivial since the fermenter sludge typically is not visually examinable. Bleeders may be partially clogged and act as filters or other phenomena such as floating or sediment layers may cause inhomogeneities. Results should therefore be checked for plausibility and possible sampling bias. Transport and storage of samples is another major source of errors. If samples can be processed within a few hours, they should be kept at process temperature in (almost filled up) closed Polyethylene (PE) or polypropylene (PP) bottles with a cannula for degassing. For longer transport/storage it depends, if DNA as the most stable, rRNA as intermediate or mRNA as the least stable nucleic acid (NA) is the target [85, 86]. In our experience, samples can be stored at ca. 4 °C for 1–2 weeks for analyses on DNA and for a few days for analyses on rRNA level. It is not finally shown for these NA species whether freezing at −20 or −80 °C respectively, and gentle thawing (at ca. 4 °C) does affect the microbial community composition. For mRNA analysis from stored samples it must be considered that this RNA species is in a highly dynamic equilibrium. Both production and degradation must be stopped immediately, e.g. by immersion in liquid N_2, acid phenol or other effective preservatives. Respective research is currently being carried out.

It must also be shown that the used NA extraction and purification system is suitable and efficient for the specific type of sample and analysis, and for quantitative analyses (see Sect. 3.1.2), the corresponding NA recovery rate has to be known. From numerous comparative studies dealing with NA extraction and purification systems it is turning out that combined physical cell disruption and chemical lysis is most suitable for environmental samples with a high portion of particulate organic matter such as fermenter sludge samples. Washing the sample prior to extraction is suggested because this substantially reduces inhibitors such as water soluble humic compounds [87]. NA purity in extracts is therefore of major concern, but as pointed out below, current guide values are not always conclusive and helpful for PCR-based assays.

Physical disruption of cells to release NAs is another crucial factor for obtaining suitable extracts. Due to velocity, ease of handling and performance efficiency, bead beating (BB) is used most frequently. Rigid cell walls must be broken, but too harsh BB can shear NAs and lead to detection failure [88]. Physical disruption must therefore be optimized for the targeted type of cells along with the particulate organic matter (OM) content in the sample sheltering the targeted cells. The higher the OM content, the more intense BB must be chosen. If differently recalcitrant cells are present, a fractionated protocol with increasing BB force and pooling of subsampled extracts can be applied [89]. It is essential to further adapt the protocol to the downstream type of analysis. If relatively short fragments such as for qPCR are suitable, relatively strong BB is of advantage. For applications requiring longer NA stretches such as functional transcriptome or genome analysis, strong BB can be counterproductive.

For RNA extracts, efficient DNase treatment and Reverse Transcription (RT) reaction with -RT controls must be performed, otherwise downstream reactions are contaminated and results biased. It must be considered that DNases degrade RNA to a certain extent, as well. This can introduce uncertainty and may only partially be overcome by method standardization leaving the possibility of a systematic error. The produced cDNA is further used just like genomic DNA but its single-stranded nature must be considered for quantitative aspects.

Downstream, extract purification is a trade-off between inhibitor removal and NA loss. PCR inhibitor removal is frequently seen as equal with matching traditional absorbance ratios (A260/230, A260/280). However, these had originally been developed for DNA-DNA hybridization and turned out to be of limited value for PCR applications. Quantitative Real-Time PCR (qPCR) was not inhibited at

A260/230 and A260/280 ratios as low as 0.02 and 1.4, respectively [32]. Cell-lysing Guanidinium-Isothiocyanat (GITC) present in some NA extraction kits absorbs at 230 nm but does not seem to compromise PCR. However, (partial) inhibition was obtained if the A320 value (humic compounds absorb at 320 nm) surpassed a level of 0.02–0.03. The A320 value thus appears to be a major indicator of PCR inhibition by samples containing humic compounds.

Between 40 % and over 80 % of extracted DNA was lost by conventional silica column post-purification [90, 91], which considerably compromises the sensitivity of quantitative assays. Optimization of the extraction/purification protocol requires that the number of treatment steps is minimized while inhibitor removal and NA recovery rates are maximized. With optimized kit-based DNA and RNA extraction/purification systems and optimized (RT)qPCR biochemistry such as inclusion of a highly processive polymerase and adjusted Mg^{2+} concentration [87], about 90 % of spiked DNA and 30–70 % of spiked viral RNA was recovered from cattle manure or biogas fermenter samples with an optimized kit-based total RNA extraction procedure [83]. However, the RNA recovery rate may actually have been lower because no DNase digestion was performed and DNase I can degrade RNA unspecifically. It is thus strongly suggested to report the method detection limit [91] of the given assay along with the DNA and/or RNA recovery rates.

3.1.2 Conventional and (Reverse-Transcription) Quantitative Real-Time PCR: Applications for Process Diagnosis

Conventional PCR is an integral step of several applications such as amplicon sequencing (see 3.1.4) and community fingerprinting (see Sect. 3.1.3). For diagnostic purposes, however, conventional PCR has lost importance in the last years in favor of Real-Time PCR (qPCR) assays. Applications, advantages and limitations of qPCR and RT-qPCR are compiled and discussed in many reviews and book chapters, e.g. [92–94]. (RT)qPCR assays avoid laborious gel-electrophoresis, are performed more quickly, are suitable for high throughput, are less prone to contamination, and provide superior specificity particularly if an additional (e.g. hydrolysis) probe or different chemistry for the same purpose is integrated [95].

The reliability of PCR assays has significantly been improved. DNA-polymerases with a very low error rate (for *Taq* ca. 3×10^{-5}, still much lower e.g. for *Pfu*, [96, 97]) and suitable reaction environment are available, and primer specificity can significantly be improved due to the enormously grown sequence data in databanks. A major issue, however, is the formation of chimaeras during PCR amplification which can seriously bias community composition analysis. Several programs and online applications can be used to check for chimaeras even in sets with relatively short amplicons [98]. Avoiding the formation of chimaeras, e.g. by analyzing templates with relatively homogenous melting temperatures (T_m) over the region of interest, would be even more straightforward than post-purifying datasets. However, it can be difficult to find template regions that provide sufficient phylogenetic resolution.

Moreover, problems or uncertainty still exist particularly with quantitative analyses of prokaryotic mRNA. This is not only due to unspecific DNase activity (see Sect. 3.1.2), but (partial) inhibition of RT-reactions that typically remain undetected. The RT efficiency at the given reaction conditions typically is not documented and may be subjected to interfering compounds introducing variability. Although these imponderabilia may not be of crucial importance for qualitative approaches such as community analyses, further methodological development is required for reliable quantification of prokaryotic mRNA.

Microbial Guilds, Bioindicators and Transcriptional Profiling

The bioindicator approach (Fig. 1) aims at analyzing and predicting distinct process states, shifts and perturbances, e.g. in biogas reactors. On the molecular level, genes encoding key enzymes of important metabolic pathways in the biogas process such as methyl-coenzyme M reductase (isogenes *mcr* and *mrt* encoding coenzyme-B sulfoethylthiotransferase, EC 2.8.4.1, the key enzyme of methanogenesis, which is present in all and exclusively in methanogenic *Archaea*), formyl-tetrahydrofolate synthetase (or formate-tetrahydrofolate ligase, *fhs*, EC 6.3.4.3, key enzyme of the Wood-Ljungdahl-pathway) or certain hydrogenases (e.g. *ech*, *hyd*) and their transcripts are ideal bioindicators and targets of molecular biomarkers [4], and even more will be identified in the near future [11, 99]. (RT)qPCR markers can be

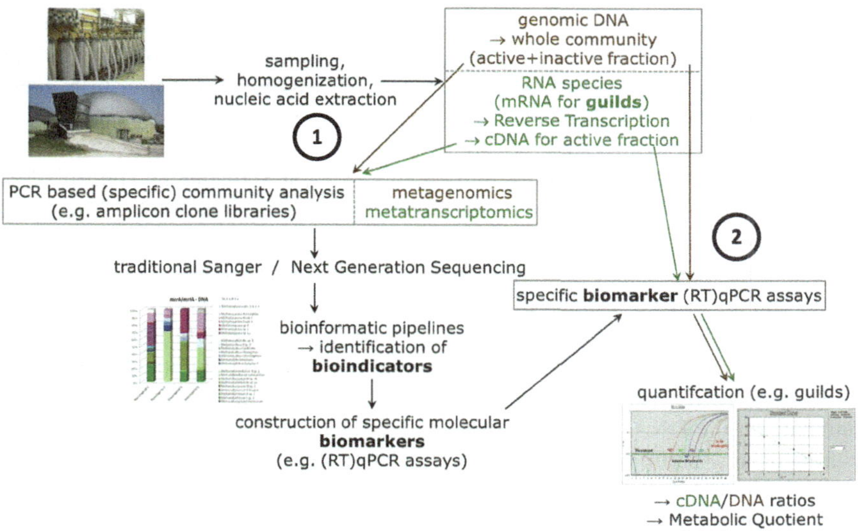

Fig. 1 The bioindicator/biomarker approach to assess process-relevant microbial guilds and their characteristic (transcriptional) activity. Once bioindicators are identified and biomarkers constructed (*1*), guilds and their (transcriptional) activity can be quantified in high-throughput assays (*2*)

tailored to determine bioindicator organisms of such guilds and their transcription activity (Fig. 1). For the design of specific biomarker systems for defined bioindicator organisms, it is essential that these had been identified previously by community composition analyses at relevant fermentation process conditions. However, the design of specific (RT)qPCR systems for heterogeneous microbial groups or clades in environmental samples can be difficult. This is particularly true if guilds are to be tracked by targeting signatures on functional genes exhibiting wobble bases in the third codon position. Respective primers, so-called "protein primers", typically are highly degenerated to provide the desired specificity, which complicates functional PCR based assays (e.g. [100]).

Due to the high stability of DNA, as compared to mRNA (see Sect. 3.1.1), DNA based assays will detect live and inactive organisms as well as residual DNA e.g. of dead organisms [101, 102]. Since for realtime process analysis, the active organisms are of particular interest, transcriptional profiling is supposed to identify more meaningful bioindicators than conventional analyses on DNA level, but changes in environmental conditions can not only induce quantitative transcription changes (see Sect. "cDNA/DNA Ratios and the Metabolic Quotient"). Typically, first metabolic activity and subsequently propagation of the populations that are best adapted to the new conditions are encouraged, whereas unadapted populations are losing competitiveness. The activity of inadequately adapted microbes is cut down first, and subsequently they are diluted out in flow-through processes. In microbial successions initiated by organic loading rate (OLR) increase in biogas processes with renewable resources, different bioindicators of the process status have been identified:

Methanosaeta spp. were present only at long microbial retention times, low acetate and ammonia levels [17, 32, 103] or/and at a feeding regime with a substrate mixture containing manure e.g. from husbandry [104, 105]. A potential bioindicator, tentatively named *Methanosaeta concilii* 2, was identified first from mesophilic maize silage digesters [4]. It is different from the *mcr*A sister clade (*M. concilii* 1) encompassing the type strain and most of the *Methanosaeta* sequences recovered from animal manure environments. Recent sequencing confirmed the presence of *M. concilii* 2 also in mesophilic grass silage digesters. This guild, probably originating from the cattle manure inoculum, soon lost transcription activity in the grass silage digestion process and was washed out subsequently (B. Munk, unpublished) at increased loading rates. *M. concilii* 2 and its activity is thus an example of a specific bioindicator of relaxed digestion conditions.

Other methanogens appear to have a similar potential to be used as indicators of the biogas process status. Results of several studies performed in different environments on the DNA and on the transcription level [4, 83, 106–108] suggest that with increasing strain to stress conditions such as shorter microbial retention times with increased SCFA concentrations and critical ammonia contents, *Methanosaetaceae* and their activity are replaced by *Methanosarcinaceae* and *Methanobacteriaceae*, with the latter appearing to be the most resistant. More specifically, certain *Methanosarcina* genospecies, hitherto undescribed *Methanosarcinaceae* (tentatively classified as genus II) and strictly hydrogenotrophic methanogens, particularly *Methanobacteriaceae*, certain *Methanobacterium* genospecies and hitherto undescribed *Methanobacteriaceae* (tentatively classified as genus IV) were increasing in maize and grass silage digestion processes on the DNA and on the transcription level [4, 83]: at aggravated strain or stress conditions, very short retention times, high SCFA or ammonia contents, at the onset of process failure, the diversity of methanogens remained almost unchanged at the DNA level, but *mcr*A/*mrt*A was transcribed exclusively by certain *Methanobacterium* genospecies (particularly *Methanobacterium* III sp. 3a) at mesophilic and *Methanothermobacter wolfeii* at thermophilic conditions. For these bioindicators, more specific (RT)qPCR based biomarker systems are being developed in order to track their presence and activity and provide a meaningful process diagnosis.

cDNA/DNA Ratios and the Metabolic Quotient

In principle, cDNA/DNA ratios can be calculated for any physiological performance of interest by relating the actual net concentration resulting from RNA transcription and transcript degradation to the concentration of the corresponding gene in a given sample. It appears to be most meaningful to determine the cDNA/DNA ratio of selected functional genes of key enzymes as activity parameter to assess the specific activity of certain guilds (see Sect. "Microbial Guilds, Bioindicators and Transcriptional Profiling", Fig. 1). Respective necessary information for designing specific (RT)qPCR system can be derived from alignments containing relevant sequences deposited in databases and extracted sequences from metagenomes and metatranscriptomes (see Sect. 3.1.4).

For *mcr*A/*mrt*A, cDNA/DNA ratios have already been reported, e.g. for peat soil and biogas fermenters [31, 109, 110]. The cDNA/DNA ratios reacted to activating stimuli such as temperature or substrate, whereas the gene concentrations remained almost constant, and they were correlated with the methane production rate within certain limits, indicating the potential of this molecular biology approach to track the activity of the guild of methanogenic *Archaea*.

Similar approaches may be envisaged to track distinct microbial activities. Concerning biogas processes, cDNA/DNA ratios, e.g. for *fhs* or *ech* subunits or other important genes of key enzymes could be very informative on the activity status of the corresponding metabolic pathways. Such information would be very helpful for process diagnosis also for e.g. biorefineries, and monitoring could provide operators with necessary information for process engineering and to decide on possible intervening measures.

However, it has to be considered that prokaryotic mRNA analysis still is delicate and error-prone, particularly if quantitative results are to be obtained (see Sects. 3.1.1 and 3.1.2). At the current state of the art, respective results should therefore be treated with precaution. RTqPCR and upstream sample preparation still need methodological development until interlaboratory comparison will create consistent and reliable results.

A second ecophysiological parameter is the Metabolic Quotient (MQ). The MQ has been developed by Munk et al. [31] and was further explained in more detail

[4]. In contrast to the entirely molecular biological parameter cDNA/DNA ratio, the MQ needs concomitant physiological data. For the MQ, the methane productivity (mL CH_4 per mL fermenter sludge) is related to the concentration of methanogenic *Archaea*, as determined by *mcr*A/*mrt*A targeted qPCR [100] on the DNA level, regardless if they are dead or alive, in the fermenter sludge at a given time, resulting in the actual specific methanogenic activity (SMA_{act}). SMA_{act} is compared to a reference standard dataset (SMA_{std}) obtained for efficient process performance at various OLRs without any symptoms of process disturbance. If SMA_{act}/SMA_{std} is >1, the methanogenic guild of interest is metabolizing at strain or stress conditions, and if SMA_{act}/SMA_{std} is <1, the methanogens are less active than at the standard reference conditions.

The MQ was measured in different maize silage digestion processes along with conventional indicators of the process state (see Sect. 2) in time series [31, 83]. It turned out that the MQ passed a threshold of about 3 ca. 2 weeks before changes were detected by the conventional chemical process indicators such as noticeable increases of the TVA/TIC ratio or SCFA concentrations. At this process stage, less methanogens than at standard conditions performed the same metabolic task, indicating metabolic strain or even stress of the given methanogenic population. When the TVA/TIC ratio and/or critical SCFAs such as propionic acid had increased to an alarming level of about 0.7 or 1 g/L, respectively, the MQ began to decline or had already decreased, indicating serious process failure and collapse. The methanogenic population was obviously seriously affected and not able anymore to fully accomplish the metabolic task of methane formation, as evidenced by the sudden decrease of methane productivity and the methane yield. When no substrate was fed to the process, the MQ was significantly below 1. According to the observations, a threshold of ca. 0.1 was defined, indicating the lowest level of normal physiologic activity.

The MQ thus allows, over the complete range of tested OLRs, to determine the metabolic state of the resident methanogenic population. A single MQ determination, however, does not necessarily mean very much. Just like with the TVA/TIC ratio, the recent development has explanatory power. An increasing MQ indicates increasing strain or stress. A decreasing MQ can indicate relaxed conditions or process breakdown. If an MQ of 1 was measured, it can be a sign of normal process operation, but it can also be symptom of a collapsed process if it turned down from values exceeding 3.

In most recent experiments with grass silage as substrate and measurements of practice biogas plants operated predominantly with grass silage, the MQ reacted similarly as in maize silage processes and stood within the bandwidth of 0.1–3 at normal process conditions without symptoms of process disturbance (B. Munk, personal communication). Since the TVA/TIC ratio is losing informative value at the high ammonia contents typically found in grass silage digestion, this is of particular importance and demonstrates the potential of the MQ as an early warning tool of process failure in practice (Fig. 1). It is expected that the MQ will become an important ecophysiological molecular microbial parameter and find application in practice monitoring of biogas plants and process diagnosis.

3.1.3 Community Fingerprinting Assays

An ideal method for microbial community analysis would allow the detection of different groups and enumerate all microbial species present in a sample from an ecosystem or habitat. Basically two approaches are used for community analyses: (1) cultivation-dependent analysis (CDA) aiming at the detection of selected groups and species of microorganisms and (2) cultivation-independent analysis (CIA), which are RNA/DNA based and are used to assess the complexity and dynamics of microbial communities. CDA relies on several selective and non-selective culture media that supply different growth conditions for specific or non-specific microbial population targeting. Traditional methods require a vast knowledge of phenotypic features to characterize microorganisms, which is often inaccurate and also leads to an underestimation of the diversity of species. However, the main drawback of conventional cultivation methods to recover less than 1 % of the total microbial species present in environmental samples remains problematic [111]. Thus CDA is nowadays complemented by molecular methods such as polymerase chain reaction (PCR) and fingerprinting techniques to assess shifts in microbial composition by small subunit ribosomal RNA gene analyses [112–114]. CIA is principally based on molecular techniques (Table 1), applying PCR and oligonucleotide probe

Table 1 Molecular screening techniques for microbial community detection, fingerprinting and identification according to [112, 119], modified

Method	Principle	Application	References
AFLP	Restriction of total microbial DNA	Strain-level identification	[120, 121]
ARDRA	Restriction of rRNA genes	Strain-level identification	[122]
ARISA	Automated riobosomal intergenic spacer region length differences (multiple ISRs per genome)	Species-level identification	[123]
RAPD	Random amplified polymorphic DNA	Strain-level identification	[124, 125]
rep-PCR	Repetitive element sequence-based PCR (short sequence repeats [SSR])	Strain-level identification	[126]
LH-PCR	Length heterogeneity of PCR-amplified rRNA genes	Community analysis	[127]
PFGE	Genomic restriction fragments in pulsed-field gel electrophoresis	Strain-level identification	[128]
T-RFLP	Terminal restriction fragment length polymorphism of rRNA genes	Strain-level identification	[129, 130]
DGGE	Mobility of partially melted dsDNA in linear gradient of DNA denaturants	Community analysis	[131,132]
TGGE	Mobility of partially melted dsDNA in linear temperature gradient	Community analysis	[133, 132]
SSCP	Mobility of conformed ssDNA in non-denaturing gels	Community analysis	[134, 135]

hybridization in order to identify microbes directly from sample material [115]. Therefore, total genomic DNA or RNA must be extracted from collected microbial cells, avoiding co-extraction of sample matrix-inherent compounds that can totally inhibit the PCR (see Sect. 3.1.1). Cell lysis is accomplished by several methods: mechanically using bead-beating, freeze-boil cycles, chemically by the use of detergents or enzymatically using cell wall degrading enzymes, e.g. lysozyme, lyticase or proteinase [116–118].

Environmental microbiological studies are often based on ribosomal DNA or RNA sequences, because these sequences are functionally and evolutionary conserved and present in all organisms. Here, 16S rDNA and 23S rDNA sequence regions have already been determined for a large number of reasonably described bacterial, archaeal and fungal species. Thus 16S rDNA sequences can be used to investigate phylogenetic relationships and for the identification of unknown microbes via comparisons with database collection entries. The largest reference databases exist for conserved marker gene 16S rRNA [112, 136]. In contrast to rDNA, rRNA targeted techniques rely on high-copy numbers per cell and are specifically used to assess changes in metabolically active microbial populations [137], although extraction and handling procedures are much more complicated due to the rRNAs instability (v. [138]). The intergenic spacer region (ISR) between 16S and 23S rDNA often shows species specific sequence variations by primers binding to conserved nucleotide stretches at the 5′ 23S and 3′ 16S rDNA gene end respectively [139]. In ISR-directed ribosomal intergenic spacer analyses (RISA) it is used to describe phylogenetic microbial diversity (*Bacteria* and *Archaea*) by creating RISA profiles. Although it is foremost used in diagnostic PCR-amplifications [140], ISR amplicons as targets for qPCR assays have also been discussed to reflect the metabolic status of key microbes more accurate than 16S based fragment comparisons [141]. Numerous broad-range and group-specific primers are available, targeting many bacterial and archaeal species of interest in AD processes, including fermentative and methanogenic representatives, covering low diversity selective cultivation sample structures up to full-scale agricultural biogas plant complex mixed community fingerprints [24, 69, 142].

The fingerprinting techniques range from simple length heterogeneity PCRs (LH-PCR) depending on different primers, targeting several variable regions in combination [143] up to more sophisticated genetic fingerprinting techniques such as amplified ribosomal DNA restriction analysis (ARDRA) or terminal restriction fragment length polymorphism (T-RFLP), which are all well established and vastly exploited to characterize whole microbial communities, providing pattern profiles of the community diversity [129, 144, 145].

T-RFLP fingerprints can give quantitative insights into communities by using a combination of fluorescence labelled primers and enzymatic digestion of resulting PCR products to generate terminal restriction fragments (T-RFs) from DNA templates. The taxonomic resolution can be improved by combining several fluorochromes and restriction enzymes simultaneously [146, 147]. Problems ascend from incomplete restriction digestion due to e.g. missing restriction sites or fragment length discrepancies caused by different fluorochromes used to estimate in silico yields, which reduces the reproducibility [130]. However, T-RFLP has not only been applied to describe bacterial communities, but also to monitor methanogenic populations and temporal shifts of archaeal communities in bioreactors [148, 149].

Further fingerprinting techniques are denaturing-gradient gel electrophoresis (DGGE), temperature-gradient gel electrophoresis (TGGE) and single-strand conformation polymorphism analysis (SSCP), which detect sequence variations of rRNA gene fragments or other functional genes from total community DNA or cDNA [150]. Complex microbial communities can thus be resolved into single members through band separation by gel electrophoresis. DNA sequence information is obtained from excised bands, which represent operational taxonomic units (OTUs) or even single species, but do often require further preparation and time consuming cloning steps. Co-mitigation or poor separation of bands representing small fragments less than 500 bp and restricted sensitivity to OTUs with a minimum abundance of 10 % (SSCP) lead to limitations of phylogenetic identifications and incomplete microbial profiles [151], which can be partly overcome by e.g. nested PCR, widening the spectrum of detectable phylogenetic groups in direct comparison to dominant members of the bacterial community [152]. In order to reach higher throughput numbers than Sanger Sequencing can provide at this stage, next generation sequencing technologies are used alternatively to analyze thousands of OTUs from different functional guilds. Yet, for complex environmental samples such as soil samples, DGGE, SSCP as well as T-RFLP provide similar compelling results on bacterial community composition [153] and microbial dynamics [152, 154], but a major drawback of ribosomal DNA based fingerprinting methods is that all DNA present in a sample is amplified, regardless the metabolic activity of bacteria, thus being less usable to reflect acute process dynamics such as crises in anaerobic digesters alone (see 3.1.2). Furthermore, molecular fingerprinting methods are not considered quantitative, but can include quantitative matrices as basis for dendrograms or can be related to multivariate analyses including process parameters, hierarchical clustering and specific microbial activities should be combined with genomic/fingerprinting data and incorporated into multivariate ordination methods such as Principal Coordinate or Principal Component Analysis (PCoA/PCA) in order to complete the whole picture drawn from a biogas biocenosis [26, 112, 155].

To evaluate environmental and process derived ecosystems, diversity is a suitable parameter that is measured by the number of different species (also from phylogenetic identification of OTUs from clone libraries assuming that one OTU corresponds to one species) including the inequality in relative abundance (Fig. 2). Therefore, diversity indices include abundance, richness and evenness as well as the Shannon index (H'). Abundance is the relative representation of a species in a community, i.e. number of a specific organism. Richness is defined as the number of different species or OTUs obtained from fingerprinting or cloning methods. Evenness is a measure of the equitability of abundance. The Shannon index (H') is calculated by the relative abundance and richness of each species (OTU or T-RFLP peak) respectively. The higher the number of phylotypes evenly distributed, the higher the H' index, which is specifically appropriate for the evaluation of low

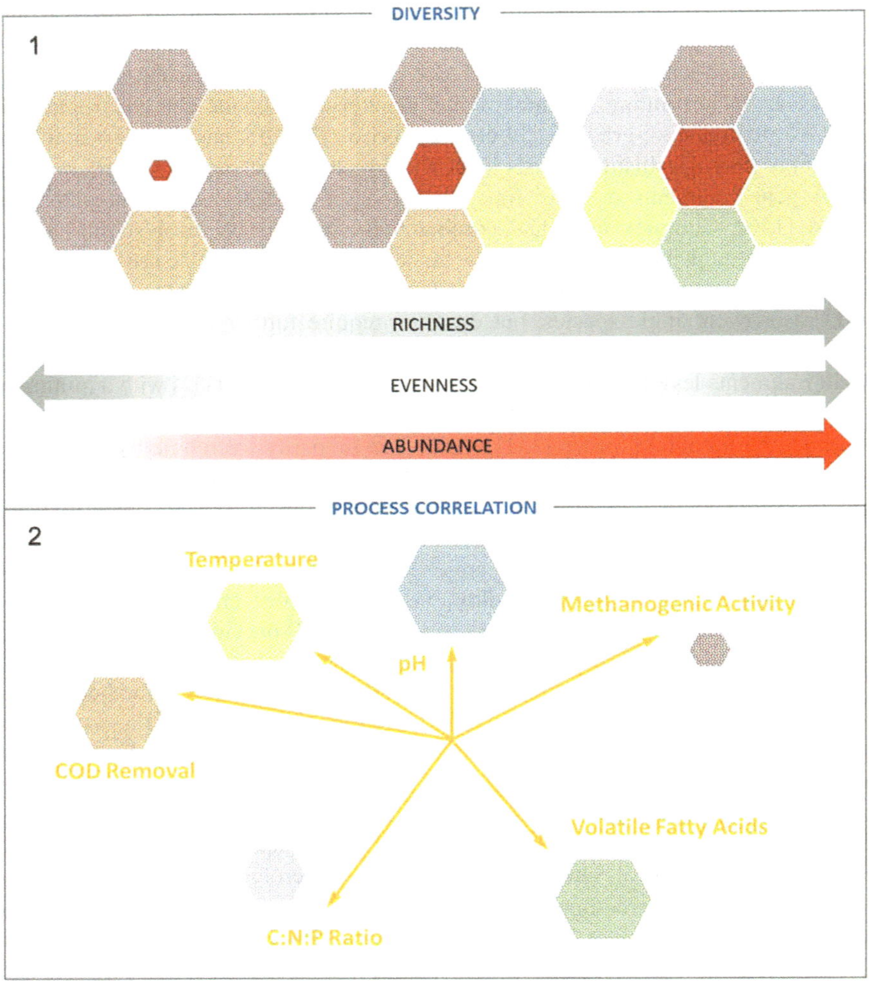

Fig. 2 Diversity—**a** complex parameter to describe the microbial composition of a given ecosystem defined by the indices richness (increases with the number of different species), evenness (distribution of present species) and abundance (number of a certain species), each colored hexagon represents one species (*1*); linking community data with process parameters, different hexagon sizes reflect the number of each species (abundances) ideal-theoretically correlated with typical process parameters and their occurrences (*2*)

abundant, but important species (indicator species) due to its high sensitivity by proportional weighting [156]. Recent community studies regarding mesophilic and thermophilic co-digestion [157] and CSTR feeding pattern comparisons [158] indicate that high initial evenness (more dynamic populations) favors the microbial functionality under selective stress conditions, suggesting the microbial community to be more flexible.

To include activity analysis, microarrays based on the hybridization of oligonucleotides or PCR products can be used to generate gene expression profiles and signatures and have been applied to investigate bacterial communities of composts as well as methanogenic communities by specifically designed microarray-chips, i.e. COMPOCHIP and ANAEROCHIP [159, 160]. This technique has also been applied in combination with real-time PCR to investigate and quantify specific targets of organic waste associated microbial communities [161]. However, the traditional microarray approach cannot detect novel genes since the device construction only involves known nucleotide sequences [162]. Therefore, metatranscriptomics described in the following section are currently preferred to enable gene expression identification without a priori sequence knowledge [162, 163] (see 3.1.4).

3.1.4 Next Generation Sequencing and Meta-Omics

The advent of affordable high-throughput Next Generation Sequencing (NGS) [84, 164] has boosted the number of sequence entries in databanks. This information has not only significantly enlarged our knowledge in systems biology; it represents an invaluable basis for further developments and exploitation. Different NGS technologies using emulsion or bridge PCR are available and can generate millions of parallel reads in small volume reactions with average read lengths between ca. 40 and about 1,100 bp. Illumina platforms are currently the most frequently used. 454 pyrosequencing will no longer be sustained. Originary Pacific Biosciences' (PacBio) RS sequencers typically generate long reads of >1 kb but the cost per base and the raw error rate (>10 % on average) are relatively high. Cost and error rates are actually lowest for Illumina and in between for Ion Torrent PGM systems [164, 165], but PacBio RS sequencing can be particularly useful e.g. by resolving problematic genomic areas such as AT-rich regions.

As compared to Sanger sequencing, error rates of these NGS systems are high, and problematic (GC-, AT-rich) regions can cause bias. Since this can result in erroneously high diversity, as observed in some ecosystem analyses, high coverage of parallel reads is required to generate reliable NGS results. Including data processing, particularly sequence assembly, all of these issues necessitate massive biocomputing efforts [166]. Respective bioinformatics pipelines and their maintenance are not affordable for any lab. Many limitations, however, will soon be overcome. Cheap annotation via cloud computing is already feasible [167], and developments towards increased read length and accuracy are going on. For example, PacBio recently introduced the RS II sequencers which are based on single molecule, real-time (SMRT) technology. PacBio claims that half of the data are in reads >14,000 base pairs with accuracy equal to Sanger sequencing. In a recent report (Mosher et al. 2014), PacBio RS II sequencing using P4/C2 chemistry surpassed the accuracy of Roche/454 pyrosequencing and generated longer reads.

For application of PCR based NGS approaches, it must additionally be considered that such amplicon sets are typically interspersed with chimaeras (see Sect. 3.1.2). These dissemble higher diversity than actually present and must be eliminated [98]. In addition, possible bias associated with Reverse Transcription or DNase treatment (see Sect. 3.1.1) must be considered [168]. However, if these challenges are adequately met, amplicon sequencing and metagenomics can produce highly comparable results, as this was shown for samples from different biogas processes by parallel analyses of curated V6-V8 (similarly as the V3-V5) 16S rDNA amplicon libraries and extracted 16S rDNA sequences from metagenomes (without interspersed selective PCR step) [169]. Although the NGS sequence numbers were much higher in this comparison, the PCR approach that was directed against the highly variable V6-V8 region provided substantially more profound insight into the bacterial community structures, occasionally even below the genus level.

NGS analysis of (complete) microbial genomes is another approach of inestimable value. It not only deepens our knowledge on microbial capacities, with the rising number of sequenced genomes and improved annotation, a more and more solid reference database is created for metagenomics and metatranscriptomics [170], leading e.g. to more reliable reference matches and improved binning accuracy. Metatranscriptomics currently is the most straightforward approach to investigate (key) metabolic pathways of interest at the transcription level, and RNAseq-based approaches allow quantitative transcriptome profiling, if suitable reference genomes or transcriptomes are available (Mutz et al. 2013). Although transcriptional activity is mostly regarded as equivalent with expression and activity, subcellular compartmentalization or excretion of enzymes and regulation are occurring at the protein level, and posttranslational modification can alter protein location and function. Additionally cross-linked metabolomics and metaproteomics might thus better reflect functional protein expression and activity in future (Vanwonterghem et al. 2014).

Although all of these approaches are providing an increasingly indispensable information background and data mining repository, they will not be applied for production scale monitoring and real-time process assays since equipment costs are too high for routine analysis, and highly skilled personnel is required. However, based on the compiled background, more meaningful and informative bioindicators may be identified and respective specifically targeted, e.g. (RT)qPCR based biomarkers could be developed (see Sect. "Microbial Guilds, Bioindicators and Transcriptional Profiling"). Such assays are much better suited for labs performing routine analyses.

3.2 *Microscopy Based Detection of Microorganisms: Specific and Non-specific Imaging*

For observations of bacterial and archaeal cells and biofilms, granules or flocs several microscopy techniques are useful, reaching from simple light microscopy (LM) with limited resolution to high-resolution scanning electron microscopy (SEM), transmission electron microscopy (TEM) and confocal laser scanning microscopy (CLSM). Amongst them, fluorescence-coupled microscopy is highly sophisticated due to its ability to detect selected groups or specific species within complex mixed communities. It is therefore widely used in microbial ecology studies allowing the visualization of spatial distribution of cells in a sample. For an in situ hybridization, a labelled probe, i.e. a fluorochromes or radioactive signal joined denatured DNA fragment is annealed to a sequence homologous to a certain target DNA (genomic DNA or PCR-amplicons). Using group- or species-specific staining, the differentiation between distinct populations is permitted leading to deep insights into the organization of biofilms and flocs [171], but strongly depends on the type of microscope used.

3.2.1 Fluorescence in Situ Hybridization-Based Confocal Laser Scanning Microscopy (FISH-CLSM)

Whereas epifluorescent imaging gives optical information from only one layer in two-dimensions [172], confocal scanning laser microscopy (CLSM) is capable of imaging a specimen via successive expositions of thin sections that can be reconstructed by computational assistance for 3D and 4D image visualization and analysis (IMARIS, Bitplane, Oxford Instruments). This allows the determination of multi-dimensional relationships of cells and their surroundings [173, 174]. The specimen is focused with a laser beam and pinhole selected fluorescent signals are detected by a photomultiplier, which results in high sensitive, high detailed and non-destructive image acquisition [175]. Fluorescence in situ hybridization targets genera or species specific ribosomal RNA fragments via probes available for *Eubacteria* (EUB) and *Archaea* (ARC). These fragments are specifically labeled with fluorescent dyes (Cy3, Cy5, FITC or FLUOS) that have individual emission wavelength optima to detect and identify multiple populations of target organisms in one sample at the same time [176]. A vast assortment of organism specific probes has already been described [177–179] and the list is constantly expanding in databases such as ‘probeBase’ [180], which provides currently over 1,300 rRNA-targeted oligonucleotide probe entries. The FISH-CLSM derived image also allows the rapid quantification of fluorescence signals, i.e. number of specific cells or percentage of area covered by biofilms [181]. Minimal statistical evaluation requires three independent samples and the observation of three individual specimen spots, when samples are homogeneous and evenly distributed [182]. At this juncture, flow cytometry (FCM) combines the advantages of microscopy and biochemical analysis for the measurement of biochemical and physical characteristics of individual cells moving in a fluid stream passing an optical sensor [177, 183, 184] (Fig. 3). In this regard, cytometric fingerprints have been reported to enable the decoding of microbial community dynamics in managed anaerobic microbial systems [185]. CLSM can also be combined with Raman spectroscopy to e.g. examine extracellular polymeric substances (EPS) producing biofilms and thereof distributed polysaccharides such as cellulose, alginate, sodium alginate, dextran, or nucleic acids during the development of the whole biofilm [186].

Numerous fluorescent dyes for DNA or RNA specific staining such as acridine orange or 4′,6-diamidino-2-phenylindole (DAPI) are used in addition to probe specific labeling, e.g. to assess the total number of bacteria against specific signals from fluorescein-labeled species [187]. Commercially available viability kits for fluorescence microscopy, e.g. Live/Dead BacLight™ (Molecular Probes®, Life Technologies) can be used to discriminate between viable and non-viable cells. Furthermore, a broad range of fluorescein-coupled molecules such as polyanionic dextrans or lectins of various molecular masses, redox-sensitive chemical probes (e.g. resofurin and fluorescein) and other fluorogenic substances (e.g. fluorescein diacetate) can be used in live cell imaging experiments to analyze (i) chemical interactions of defined molecules, (ii) cellular physiological conditions about

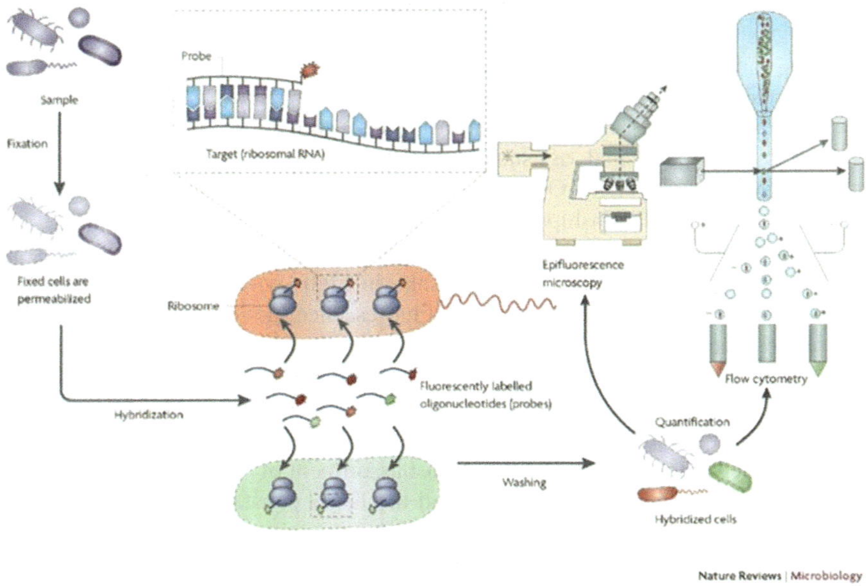

Fig. 3 Single-cell identification and quantification by either epifluorescence microscopy or quantitative flow cytometry on the basis of fluorescence in situ hybridization according to Amann and Fuchs [177]. The sample preparation involves the fixation of microbial cells to stabilize and permeabilize their membranes to allow labelled oligonucleotide probes to access and hybridize to certain intracellular targets. Adapted from Jul 20, 2015, Nature Publishing Group

membrane potential or permeability and (iii) microzonal variations in biofilm chemistry regarding pH, redox potential or ion concentrations [188–190]. FISH is also performed in combination with fingerprinting methods (see Sect. 3.1.3) or cloning experiments as full-cycle rRNA approach to quantitatively determine the relevance and spatial distribution of given operational taxonomic units (OTUs) [191].

Microautoradiography-coupled FISH (MAR-FISH) is another tool for structure and function analyses in microbial ecology [192] that links phylotypic characteristics with metabolic activities to reveal microbial species responsible for key physiological processes [193, 194]. The microbial in situ uptake and incorporation of radioactively labelled substrates can be visualized and enumerated this way [195], but the method is limited to elements with radioactive isotopes (e.g. $^{13/14}C$, ^{3}H, ^{15}N, ^{34}S, ^{33}P, ^{18}O), which makes secondary ion mass spectrometry (SIMS) become a constitutive alternative for MAR-FISH. However, in anaerobic digesters it has been used to elucidate metabolic functions of minor phylogenetic groups like *Chloroflexi*, *Syntrophomonas Spirochaeta* and *Synergistes* as well as *Methanosaeta* spp. in sugar and short fatty acid such as acetate, butyrate, and propionate utilization [196, 197], and led to the determination of degradation rates of glucose, acetate and propionate as well [198, 199]. Although not providing quantitative data, the complementary combination of MAR-FISH with quantitative real-time PCR can be useful to investigate active key functional microbial groups [112].

FISH probing and CLSM have thus been used to show, how microbial communities involved in the anaerobic biodegradation process are organized regarding biofilm formation, immobilization and attachment to solid substrate material [64, 200–202], but also to study bioreactor and full-scale biogas plant performances [203–205]. There are some drawbacks using hybridization based fluorescence microscopy that include fading or photo-bleaching of the fluorochromes, fluorescence quenching, the loss of fluorescence due to sample derived molecules interacting with the fluorochromes, limited archaeal cell wall permeability and inefficient or incorrect hybridization. Many of these problems can be overcome by modifications of the preparation protocol towards sample and organism (i.e. gram-positive/gram-negative *Bacteria* and methanogenic *Archaea*) optimized hybridization conditions such as temperature or formamide concentrations [206], enzymatic pretreatment as routinely applied for catalyzed reporter deposition FISH (CARD-FISH) [207], up to double labeling (DOPE-FISH) for improved signal intensity and rRNA accessibility [208, 209], or even individual probe design [177, 210]. FISH-CLSM is clearly a valuable technique for AD processes to analyze microbial dynamics, since both, qualitative and quantitative information can be obtained, but specialized personnel and laboratory equipment is required to perform these analyses.

3.2.2 High-Resolution Microscopy: Scanning Electron Microscopy (SEM)

Apart from molecular biology depending light microscopy techniques, biofilms and single cells can be also investigated by scanning electron microscopy with unequalled magnifications of up to 500,000-fold (Carl Zeiss Ultra 55, Hitachi S-3000 N). New microscope-generations like RISA even integrate correlative, confocal Raman imaging with scanning electron microscopy (Raman-SEM), permitting a direct link between ultra-structural surface properties and molecular compound information (WITec, TESCAN). Therefore, scanning and transmission electron microscopy (SEM/TEM) have not only been used to study overall biofilm organization patterns (see below), but to investigate cell-to-cell interactions of anaerobic digestion process innate syntrophic microbial partners on a nano-scale level such as the interspecies electron transfer [211]. Direct interspecies electron transfer (DIET) depends on hydrogen and carbon source such as ethanol or formate, which was recently discovered for *Geobacter metallireducens* and *Methanosaeta harundinacea* or *Methanosarcina barkeri* interactions to lastly reduce carbon dioxide to methane [212]. It has also been shown, that growth of fermentative and methanogenic microbes on conductive carriers is tangible, suggesting *Bacteria* (most likely Clostriaceae) and methanogenic *Archaea* (most likely Methanobacteriaceae) can transfer electrons from a stainless steel support even without the involvement of hydrogen or formate [213].

Considering the low growth rate of methanogenic *Archaea*, immobilization on support material such as polymers (e.g. polyurethane, acrylonitrile-acrylamide, nylon) is a potential strategy to allow longer residence times in bioreactors for the adjustment to unstable conditions and varying feeding regimes [214] as shown by SEM for lab-scale reactors continuously operated with vinasse waste to keep COD removal rates constant at decreased retention times and various organic loading rates [215]. SEM was also used to study the natural biofilm formation on zeolite particles during *in sacco* incubation in semi-continuously, completely stirred lab-scale fermenters fed with grass silage [64, 202], or comparing several other carriers for *Bacteria* and *Archaea* such as activated carbon, polyvinyl alcohol or glass fibers in anaerobic digesters treating cattle manure [216], demonstrating that specific materials can selectively support methanogens to avoid co-cultivation of unwanted sulfate-reducing bacteria (SRB) during anaerobic wastewater treatment and methane production from molasses [217]. Focusing feedstock for AD, SEM can be part of efficiency evaluations of pretreatment methods for specific substrates such as that steam explosion induces significant morphological changes in treated lignocellulosic materials [218, 219].

In addition to the direct observation of sputtered organic matter, energy dispersive X-ray spectroscopy (SEM-EDS/EDX) allows element analyses of inorganic sample components of carrier materials or to characterize stable and active catalysts for hydrogen production from biogas, using SEM-TEM in combination with other microscopy methods to evaluate the deposition or arrangement of hollow carbon nanotubes and nanofibers [220]. Furthermore, EDS and TEM can be used for the localization of substrates or electron donors and acceptors or characterization of e.g. metal transformation in metal-reducing bacteria. However, the major drawback of electron microscopy is that it is an invasive method, which requires sample fixation and preparation including consecutive dehydration steps for specimen observations in high vacuum. Biological structures can be maintained by critical point drying, lyophilisation or high-pressure freezing. Instead of SEM, environmental SEM using lower vacuum pressures can be used alternatively as well. Pinpoint extraction and ultra-thin layer observations by consecutive cryosectioning are further techniques to investigate certain regions of interest and cellular aspects respectively based on SEM/TEM or focused ion beam (FIB)-SEM that can be also combined with CLSM 3D imaging for real 3D correlations of one and the same biological event in an identical sample [221, 222].

References

1. Schink B (2006) Syntrophic associations in methanogenic degradation. In: Molecular basis of symbiosis, vol 41. Springer, Berlin, pp 1–19
2. McInerney MJ, Struchtemeyer CG, Sieber JR, Mouttaki H, Stams AJM, Schink B, Rohlin L, Gunsalus RP (2008) Physiology, ecology, phylogeny, and genomics of microorganisms capable of syntrophic metabolism. Ann N Y Acad Sci 1125:58–72
3. Demirel B, Scherer P (2008) The roles of acetotrophic and hydrogenotrophic methanogens during anaerobic conversion of biomass to methane: a review. Rev Environ Sci Biotechnol 7 (2):173–190
4. Lebuhn M, Munk B, Effenberger M (2014) Agricultural biogas production in Germany—from practice to microbiology basics. Energy Sustain Soc 4(10):21
5. Thauer RK, Kaster A-K, Seedorf H, Buckel W, Hedderich R (2008) Methanogenic archaea: ecologically relevant differences in energy conservation. Nat Rev Microbiol 6(8):579–591
6. Stams AJM, Oude Elferink SJWH, Westermann P (2003) Metabolic interactions between methanogenic consortia and anaerobic respiring bacteria. Adv Biochem Eng Biotechnol 81:31–56
7. Fotidis IA, Karakashev D, Angelidaki I (2014) The dominant acetate degradation pathway/methanogenic composition in full-scale anaerobic digesters operating under different ammonia levels. Int J Environ Sci Technol 11(7):2087–2094
8. Schink B (1997) Energetics of syntrophic cooperation in methanogenic degradation. Microbiol Mol Biol Rev 61(2):262–280
9. Stams AJM, Plugge CM (2009) Electron transfer in syntrophic communities of anaerobic bacteria and archaea. Nat Rev Microbiol 7(8):568–577
10. McInerney MJ, Sieber JR, Gunsalus RP (2009) Syntrophy in anaerobic global carbon cycles. Curr Opin Biotechnol 20(6):623–632
11. Worm P, Koehorst JJ, Visser M, Sedano-Núñez VT, Schaap PJ, Plugge CM, Sousa DZ, Stams AJM (2014) A genomic view on syntrophic versus non-syntrophic lifestyle in anaerobic fatty acid degrading communities. Biochim Biophys Acta—Bioenerg 1837 (12):2004–2016
12. Solli L, Håvelsrud OE, Horn SJ, Rike AG (2014) A metagenomic study of the microbial communities in four parallel biogas reactors. Biotechnol Biofuels 7(1):146
13. Kazda M, Langer S, Bengelsdorf FR (2014) Fungi open new possibilities for anaerobic fermentation of organic residues. Energy Sustain Soc 4(1):6

14. Koeck DE, Wibberg D, Maus I, Winkler A, Albersmeier A, Zverlov VV, Liebl W, Pühler A, Schwarz WH, Schlüter A (2014) Complete genome sequence of the cellulolytic thermophile Ruminoclostridium cellulosi wild-type strain {DG}5 isolated from a thermophilic biogas plant. J Biotechnol 188:136–137
15. Gruninger RJ, Puniya AK, Callaghan TM, Edwards JE, Youssef N, Dagar SS, Fliegerova K, Griffith GW, Forster R, Tsang A, McAllister T, Elshahed MS (2014) Anaerobic fungi (phylum Neocallimastigomycota): advances in understanding their taxonomy, life cycle, ecology, role and biotechnological potential. FEMS Microbiol Ecol 90(1):1–17
16. Weiland P (2010) Biogas production: current state and perspectives. Appl Microbiol Biotechnol 85(4):849–860
17. Nikolausz M, Walter RFH, Sträuber H, Liebetrau J, Schmidt T, Kleinsteuber S, Bratfisch F, Günther U, Richnow HH (2013) Evaluation of stable isotope fingerprinting techniques for the assessment of the predominant methanogenic pathways in anaerobic digesters. Appl Microbiol Biotechnol 97(5):2251–2262
18. Lv Z, Hu M, Harms H, Richnow HH, Liebetrau J, Nikolausz M (2014) Stable isotope composition of biogas allows early warning of complete process failure as a result of ammonia inhibition in anaerobic digesters. Bioresour Technol 167:251–259
19. Polag D, Krapf LC, Heuwinkel H, Laukenmann S, Lelieveld J, Keppler F (2014) Stable carbon isotopes of methane for real-time process monitoring in anaerobic digesters. Eng Life Sci 14(2):153–160
20. Van Soest PJ, McQueen RW (1973) The chemistry and estimation of fibre. Proc Nutr Soc 32 (3):123–130
21. Krapf LC, Heuwinkel H, Schmidhalter U, Gronauer A (2013) The potential for online monitoring of short-term process dynamics in anaerobic digestion using near-infrared spectroscopy. Biomass Bioenergy 48:224–230
22. da Silva TL, Roseiro JC, Reis A (2012) Applications and perspectives of multi-parameter flow cytometry to microbial biofuels production processes. Trends Biotechnol 30(4):225–232
23. Rosselló-Móra R (2012) Towards a taxonomy of Bacteria and Archaea based on interactive and cumulative data repositories. Environ Microbiol 14:318–334
24. Stantscheff R, Kuever J, Rabenstein A, Seyfarth K, Dröge S, König H (2014) Isolation and differentiation of methanogenic Archaea from mesophilic corn-fed on-farm biogas plants with special emphasis on the genus Methanobacterium. Appl Microbiol Biotechnol 98 (12):5719–5735
25. Musat N, Foster R, Vagner T, Adam B, Kuypers MMM (2012) Detecting metabolic activities in single cells, with emphasis on nanoSIMS. FEMS Microbiol Rev 36:486–511
26. Vanwonterghem I, Jensen PD, Ho DP, Batstone DJ, Tyson GW (2014) Linking microbial community structure, interactions and function in anaerobic digesters using new molecular techniques. Curr Opin Biotechnol 27:55–64
27. Voß E, Weichgrebe D, Rosenwinkel KH (2009) FOS/TAC: Herleitung, Methodik, Anwendung und Aussagekraft. In: Proceedings of the international scientific conference biogas science 2009, vol 3, pp 675–682
28. Quist-Wessel PMF, Langeveld J (2014) Bioenergy production from waste agricultural biomass. In: Langeveld H, Dixon J, Van Keulen H (eds) Biofuel cropping systems carbon, land and food. Taylor and Francis, Hoboken
29. Yenigün O, Demirel B (2013) Ammonia inhibition in anaerobic digestion: a review. Process Biochem 48(5–6):901–911
30. Hansen KH, Angelidaki I, Ahring BK (1998) Anaerobic digestion of swine manure: inhibition by ammonia. Water Res 32(1):5–12
31. Munk B, Bauer C, Gronauer A, Lebuhn M (2011) A metabolic quotient for methanogenic Archaea. Water Sci Technol 66(11):2311–2317
32. Munk B, Bauer C, Gronauer A, Lebuhn M (2010) Population dynamics of methanogens during acidification of biogas fermenters fed with maize silage. Eng Life Sci 10(6):496–508

33. Demirel B, Scherer P (2011) Trace element requirements of agricultural biogas digesters during biological conversion of renewable biomass to methane. Biomass Bioenergy 35 (3):992–998
34. Vintiloiu A, Lemmer A, Oechsner H, Jungbluth T (2012) Mineral substances and macronutrients in the anaerobic conversion of biomass: an impact evaluation. Eng Life Sci 12(3):287–294
35. Chen Y, Cheng JJ, Creamer KS (2008) Inhibition of anaerobic digestion process: a review. Bioresour Technol 99(10):4044–4064
36. Mudhoo A, Kumar S (2013) Effects of heavy metals as stress factors on anaerobic digestion processes and biogas production from biomass. Int J Environ Sci Technol 10(6):1383–1398
37. Hecht C, Griehl C (2009) Investigation of the accumulation of aromatic compounds during biogas production from kitchen waste. Bioresour Technol 100(2):654–658
38. Sayder B, Vitz H, Mohring S, Merrettig-Bruns U, Kabasci S, Hamscher G, Tuerk J (2009) Gehemmte Biologie. Biogas-J 12(2):44–45
39. Young JC (2004) Respirometry for environmental science and engineering. SJ Enterprises, Springdale, Arkansas
40. Noike T, Endo G, Chang JE, Yaguchi J, Matsumoto J (1985) Characteristics of carbohydrate degradation and the rate-limiting step in anaerobic digestion. Biotechnol Bioeng 27 (10):1482–1489
41. Lynd LR, Weimer PJ, van Zyl WH, Pretorius IS (2002) Microbial cellulose utilization: fundamentals and biotechnology. Microbiol. Mol Biol Rev 66,(3):506–577
42. Merlin Christy P, Gopinath LR, Divya D (2014) A review on anaerobic decomposition and enhancement of biogas production through enzymes and microorganisms. Renew Sustain Energy Rev 34(C):167–173
43. Arantes V, Saddler JN (2010) Access to cellulose limits the efficiency of enzymatic hydrolysis: the role of amorphogenesis. Biotechnol Biofuels 3:4
44. Zhang Y-HP, Lynd LR (2005) Determination of the number-average degree of polymerization of cellodextrins and cellulose with application to enzymatic hydrolysis. Biomacromolecules 6(3):1510–1515
45. Schwald W, Chart M, Breuil C, Saddler JN (1988) Applied microbiology biotechnology comparison of HPLC and colorimetric methods for measuring cellulolytic activity, pp 398–403
46. Khan AW, Tremblay D, LeDuy A (1986) Assay of xylanase and xylosidase activities in bacterial and fungal cultures. Enzyme Microb Technol 8(6):373–377
47. Obst U (1985) Test instructions for measuring the microbial metabolic activity in water samples. Fresenius' Zeitschrift fuer Anal Chemie 321(2):166–168
48. Gasch C, Hildebrandt I, Rebbe F, Röske I (2013) Enzymatic monitoring and control of a two-phase batch digester leaching system with integrated anaerobic filter. Energy Sustain Soc 3(1):10
49. Mshandete AM, Björnsson L, Kivaisi AK, Steven M, Rubindamayugi T, Mattiasson B (2008) Two-stage anaerobic digestion of aerobic pre-treated sisal leaf decortications residues: hydrolases activities and biogas production profile. Afr J Biochem Res 2(5):211–218
50. Holzapfel-Pschorn A, Obst U, Haberer K (1987) Sensitive methods for the determination of microbial activities in water samples using fluorigenic substrates. Fresenius' Zeitschrift fuer Anal Chemie 327(5–6):521–523
51. Regueiro L, Veiga P, Figueroa M, Alonso-Gutierrez J, Stams AJM, Lema JM, Carballa M (2012) Relationship between microbial activity and microbial community structure in six full-scale anaerobic digesters. Microbiol Res 167(10):581–589
52. Parawira W (2012) Enzyme research and applications in biotechnological intensification of biogas production. Crit Rev Biotechnol 32:172–186
53. Ahuja SK, Ferreira GM, Moreira AR (2004) Utilization of enzymes for environmental applications. Crit Rev Biotechnol 24(2–3):125–154

54. Warthmann R, Baier U, Meier U, Hersener J (2013) Optimisation of anaerobic digestion by pre-treatment, additives and process engineering. In: Biogas process optimisation workshop, 2013
55. Bruni E, Jensen AP, Angelidaki I (2010) Comparative study of mechanical, hydrothermal, chemical and enzymatic treatments of digested biofibers to improve biogas production. Bioresour Technol 101(22):8713–8717
56. Shen S, Nges IA, Yun J, Liu J (2014) Pre-treatments for enhanced biochemical methane potential of bamboo waste. Chem Eng J 240:253–259
57. Yu S, Zhang G, Li J, Zhao Z, Kang X (2013) Effect of endogenous hydrolytic enzymes pretreatment on the anaerobic digestion of sludge. Bioresour Technol 146:758–761
58. Diak J, Örmeci B, Kennedy KJ (2012) Effect of enzymes on anaerobic digestion of primary sludge and septic tank performance. Bioprocess Biosyst Eng 35(9):1577–1589
59. Plöchl M, Hilse A, Heiermann M, Suarez Quinones T, Budde J, Prochnow A (2009) Application of hydrolytic enzymes for improving biogas feedstock fluidity. CIGR E J IX:1–16
60. Burgess J, Pletschke B (2008) Hydrolytic enzymes in sewage sludge treatment: a mini-review. Water Sa 34(3):343–350
61. Davidsson Å, Wawrzynczyk O Norrlöw, Jansen J la C (2007) Strategies for enzyme dosing to enhance anaerobic digestion of sewage sludge. J Residuals Sci Technol 4(1):1–7
62. Vavilin VA, Rytov SV, Lokshina LY (1996) A description of hydrolysis kinetics in anaerobic degradation of particulate organic matter. Bioresour Technol 56(2–3):229–237
63. Gokhale A, Lu J, Lee I (2013) Immobilization of cellulase on magnetoresponsive graphene nano-supports. J Mol Catal B Enzym 90:76–86
64. Weiss S, Lebuhn M, Andrade D, Zankel A, Cardinale M, Birner-Gruenberger R, Somitsch W, Ueberbacher BJ, Guebitz GM (2013) Activated zeolite—Suitable carriers for microorganisms in anaerobic digestion processes? Appl Microbiol Biotechnol 97(7):3225–3238
65. Montalvo S, Guerrero L, Borja R, Sánchez E, Milán Z, Cortés I, Angeles de la la Rubia M (2012) Application of natural zeolites in anaerobic digestion processes: a review. Appl Clay Sci 58:125–133
66. Jeganathan J, Nakhla G, Bassi A (2007) Hydrolytic pretreatment of oily wastewater by immobilized lipase. J Hazard Mater 145(1–2):127–135
67. Kavitha S, Adish Kumar S, Yogalakshmi KN, Kaliappan S, Rajesh Banu J (2013) Effect of enzyme secreting bacterial pretreatment on enhancement of aerobic digestion potential of waste activated sludge interceded through EDTA. Bioresour Technol 150:210–219
68. Wang W, Yan L, Cui Z, Gao Y, Wang Y, Jing R (2011) Characterization of a microbial consortium capable of degrading lignocellulose. Bioresour Technol 102(19):9321–9324
69. Yan L, Gao Y, Wang Y, Liu Q, Sun Z, Fu B, Wen X, Cui Z, Wang W (2012) Diversity of a mesophilic lignocellulolytic microbial consortium which is useful for enhancement of biogas production. Bioresour Technol 111:49–54
70. Weiss S, Tauber M, Somitsch W, Meincke R, Müller H, Berg G, Guebitz GM (2010) Enhancement of biogas production by addition of hemicellulolytic bacteria immobilised on activated zeolite. Water Res 44(6):1970–1980
71. Kovács KL, Ács N, Kovács E, Wirth R, Rákhely G, Strang O, Herbel Z, Bagi Z (2013) Improvement of biogas production by bioaugmentation. Biomed Res Int 2013:482653
72. Bagi Z, Acs N, Bálint B, Horváth L, Dobó K, Perei KR, Rákhely G, Kovács KL (2007) Biotechnological intensification of biogas production. Appl Microbiol Biotechnol 76(2):473–482
73. Peng X, Börner RA, Nges IA, Liu J (2014) Impact of bioaugmentation on biochemical methane potential for wheat straw with addition of Clostridium cellulolyticum. Bioresour Technol 152:567–571
74. Levine SE, Fox JM, Blanch HW, Clark DS (2010) A mechanistic model of the enzymatic hydrolysis of cellulose. Biotechnol Bioeng 107(1):37–51
75. Christ O, Wilderer PA, Angerhöfer R, Faulstich M (2000) Mathematical modeling of the hydrolysis of anaerobic processes. Water Sci Technol 41(3):61–65

76. Rajendran K, Kankanala HR, Lundin M, Taherzadeh MJ (2014) A novel process simulation model (PSM) for anaerobic digestion using Aspen Plus. Bioresour Technol 168:7–13
77. Koch K, Lübken M, Gehring T, Wichern M, Horn H (2010) Biogas from grass silage—Measurements and modeling with ADM1. Bioresour Technol 101(21):8158–8165
78. Ramirez I, Mottet A, Carrère H, Déléris S, Vedrenne F, Steyer J-P (2009) Modified ADM1 disintegration/hydrolysis structures for modeling batch thermophilic anaerobic digestion of thermally pretreated waste activated sludge. Water Res 43(14):3479–3492
79. Batstone D (2001) The IWA anaerobic digestion model no 1, vol 1, no. 1, pp 1–68
80. Narihiro T, Kamagata Y (2013) Cultivating Yet-to-be Cultivated Microbes: The Challenge Continues. Microbes Environ 28(2):163–165
81. Hugenholtz P (2002) Exploring prokaryotic diversity in the genomic era. Genome Biol 3 (2):0003.1–0003.8
82. Westerholm M, Levén L, Schnürer A (2012) Bioaugmentation of syntrophic acetate-oxidizing culture in biogas reactors exposed to increasing levels of ammonia. Appl Environ Microbiol 78(21):7619–7625
83. Munk B, Lebuhn M (2014) Process diagnosis using methanogenic Archaea in maize-fed, trace element depleted fermenters. Anaerobe 29:22–28
84. Metzker ML (2010) Sequencing technologies—the next generation. Nat Rev Genet 11 (1):31–46
85. Fontaine M, Guillot E (2003) Study of 18S rRNA and rDNA stability by real-time RT-PCR in heat-inactivated Cryptosporidium parvum oocysts. FEMS Microbiol Lett 226(2):237–243
86. Keer JT, Birch L (2003) Molecular methods for the assessment of bacterial viability. J Microbiol Methods 53(2):175–183
87. Lebuhn M, Effenberger M, Gronauer A, Wilderer PA, Wuertz S (2003) Using quantitative real-time PCR to determine the hygienic status of cattle manure. Water Sci Technol 48 (4):97–103
88. Garcés G (2005) Quantitative real-time PCR for detecting Cryptosporidium parvum in cattle manure and anaerobic digester samples—Methodological advances in {DNA} extraction. In: Proceedings of the 8th Latin American workshop and symposium on anaerobic digestion, pp 68–73
89. Garcés-Sanchez G, Wilderer PA, Munch JC, Horn H, Lebuhn M (2009) Evaluation of two methods for quantification of hsp70 mRNA from the waterborne pathogen Cryptosporidium parvum by reverse transcription real-time PCR in environmental samples. Water Res 43 (10):2669–2678
90. Schmedes S, Marshall P, King JL, Budowle B (2013) Effective removal of co-purified inhibitors from extracted DNA samples using synchronous coefficient of drag alteration (SCODA) technology. Int J Legal Med 127(4):749–755
91. Lebuhn M, Effenberger M, Garcés G, Gronauer A, Wilderer PA (2004) Evaluating real-time PCR for the quantification of distinct pathogens and indicator organisms in environmental samples. Water Sci Technol 50(1):263–270
92. Dorak MT (ed) (2007) Real-time PCR, 1st edn. Taylor & Francis, New York
93. Lech M, Anders H-J (2014) Expression profiling by real-time quantitative polymerase chain reaction (RT-qPCR). Methods Mol Biol 1169:133–142
94. Kim J, Lim J, Lee C (2013) Quantitative real-time PCR approaches for microbial community studies in wastewater treatment systems: Applications and considerations. Biotechnol Adv 31 (8):1358–1373
95. Navarro E, Serrano-Heras G, Castaño MJ, Solera J (2015) Real-time PCR detection chemistry. Clin Chim Acta 439:231–250
96. Terpe K (2013) Overview of thermostable DNA polymerases for classical PCR applications: from molecular and biochemical fundamentals to commercial systems. Appl Microbiol Biotechnol 97(24):10243–10254
97. McInerney P, Adams P, Hadi MZ (2014) Error rate comparison during polymerase chain reaction by DNA Polymerase. Mol Biol Int 2014:8

98. Haas BJ, Gevers D, Earl AM, Feldgarden M, Ward DV, Giannoukos G, Ciulla D, Tabbaa D, Highlander SK, Sodergren E, Methé B, DeSantis TZ (2011) Human microbiome consortium. In: F. Petrosino J, Knight R, Birren BW (eds) Chimeric 16S rRNA sequence formation and detection in Sanger and 454-pyrosequenced PCR amplicons. Genome Res 21(3):494–504
99. Sieber JR, McInerney MJ, Gunsalus RP (2012) Genomic insights into syntrophy: the paradigm for anaerobic metabolic cooperation. Annu Rev Microbiol 66(1):429–452
100. Bauer C, Korthals M, Gronauer A, Lebuhn M (2008) Methanogens in biogas production from renewable resources—a novel molecular population analysis approach. Water Sci Technol 58(7):1433–1439
101. Widmer G, Orbacz EA, Tzipori S (1999) beta-tubulin mRNA as a marker of Cryptosporidium parvum oocyst viability. Appl Environ Microbiol 65(4):1584–1588
102. McKillip JL, Jaykus LA, Drake M (1999) Nucleic acid persistence in heat-killed Escherichia coli O157:H7 from contaminated skim milk. J Food Prot 62(8):839–844
103. Ho DP, Jensen PD, Batstone DJ (2013) Methanosarcinaceae and acetate-oxidizing pathways dominate in high-rate thermophilic anaerobic digestion of waste-activated sludge. Appl Environ Microbiol 79(20):6491–6500
104. Nettmann E, Bergmann I, Pramschüfer S, Mundt K, Plogsties V, Herrmann C, Klocke M (2010) Polyphasic analyses of methanogenic archaeal communities in agricultural biogas plants. Appl Environ Microbiol 76(8):2540–2548
105. Li J, Rui J, Pei Z, Sun X, Zhang S, Yan Z, Wang Y, Liu X, Zheng T, Li X (2014) Straw- and slurry-associated prokaryotic communities differ during co-fermentation of straw and swine manure. Appl Microbiol Biotechnol 98(10):4771–4780
106. Steinberg LM, Regan JM (2011) Response of lab-scale methanogenic reactors inoculated from different sources to organic loading rate shocks. Bioresour Technol 102:8790–8798
107. Zhang C, Yuan Q, Lu Y (2014) Inhibitory effects of ammonia on methanogen mcrA transcripts in anaerobic digester sludge. FEMS Microbiol Ecol 87(2):368–377
108. Lv Z, Leite AF, Harms H, Richnow HH, Liebetrau J, Nikolausz M (2014) Influences of the substrate feeding regime on methanogenic activity in biogas reactors approached by molecular and stable isotope methods. Anaerobe 29:91–99
109. Freitag TE, Prosser JI (2009) Correlation of methane production and functional gene transcriptional activity in a peat soil. Appl Environ Microbiol 75(21):6679–6687
110. Watanabe T, Kimura M, Asakawa S (2009) Distinct members of a stable methanogenic archaeal community transcribe mcrA genes under flooded and drained conditions in Japanese paddy field soil. Soil Biol Biochem 41(2):276–285
111. Amann RI, Ludwig W, Schleifer KH (1995) Phylogenetic identification and in situ detection of individual microbial cells without cultivation. Microbiol Rev 59(1):143–169
112. Talbot G, Topp E, Palin MF, Massé DI (2008) Evaluation of molecular methods used for establishing the interactions and functions of microorganisms in anaerobic bioreactors. Water Res 42(3):513–537
113. Rolfs A, Schuller I, Finckh U, Weber-Rolfs I (1992) Detection of single base changes using PCR. In: Rolfs A, Schuller I, Finckh U, Weber-Rolfs I (eds) PCR: clinical diagnostics and research. Springer, Berlin, pp 149–167
114. Mullis K, Faloona F, Scharf S, Saiki R, Horn G, Erlich H (1986) Specific enzymatic amplification of DNA in vitro: the polymerase chain reaction. Cold Spring Harb Symp Quant Biol 51:263–273
115. Frigon D, Oerthe DB, Morgenroth E, Raskin L (2002) Oligonucleotide probe hybridization and modeling results suggest that populations consuming readily degradable substrate have high cellular RNA levels. Water Sci Technol 45(6):115–126
116. Candrian U (1995) Polymerase chain reaction in food microbiology. J Microbiol Methods 23 (1):89–103
117. Roose-Amsaleg C, Garnier-Sillam E, Harry M (2001) Extraction and purification of microbial DNA from soil and sediment samples. Appl. Soil Ecol 18(1):47–60

118. Griffiths RI, Whiteley AS, O'Donnell AG, Bailey MJ (2000) Rapid method for coextraction of DNA and RNA from natural environments for analysis of ribosomal DNA- and rRNA-based microbial community composition. Appl Environ Microbiol 66(12):5488–5491
119. Maukonen J, Mättö J, Wirtanen G, Raaska L, Mattila-Sandholm T, Saarela M (2003) Methodologies for the characterization of microbes in industrial environments: a review. J Ind Microbiol Biotechnol 30(6):327–356
120. Vos P, Hogers R, Bleeker M, Reijans M, van de Lee T, Hornes M, Frijters A, Pot J, Peleman J, Kuiper M (1995) AFLP: a new technique for DNA fingerprinting. Nucleic Acids Res 23(21):4407–4414
121. Janssen P, Bleeker M, Kersters K (1996) Evaluation of the DNA fingerprinting method AFLP as a new tool in bacterial taxonomy, pp 1881–1893
122. Sklarz MY, Angel R, Gillor O, Soares MIM (2009) Evaluating amplified rDNA restriction analysis assay for identification of bacterial communities. Antonie Van Leeuwenhoek 96 (4):659–664
123. Fisher MM, Triplett EW (1999) Automated approach for ribosomal intergenic spacer analysis of microbial diversity and its application to freshwater bacterial communities. Appl Environ Microbiol 65(10):4630–4636
124. Welsh J, McClelland M (1990) Fingerprinting genomes using PCR with arbitrary primers. Nucleic Acids Res 18(24):7213–7218
125. Power EG (1996) RAPD typing in microbiology—a technical review. J Hosp Infect 34 (4):247–265
126. Versalovic J, Schneider M, de Bruijn FJ, Lupski JR (1994) Genomic fingerprinting of bacteria using repetitive sequence-based polymerase chain reaction. Methods Mol Cell Biol 5 (25):25–40
127. Mills DK, Entry JA, Gillevet PM, Mathee K (2007) Assessing microbial community diversity using amplicon length heterogeneity polymerase chain reaction. Soil Sci Soc Am J 71(2):572
128. Schwartz DC, Cantor CR (1984) Separation of yeast chromosome-sized DNAs by pulsed field gradient gel electrophoresis. Cell 37(1):67–75
129. Liu WT, Marsh TL, Cheng H, Forney LJ (1997) Characterization of microbial diversity by determining terminal restriction fragment length polymorphisms of genes encoding 16S rRNA. Appl Environ Microbiol 63(11):4516–4522
130. Nocker A, Burr M, Camper AK (2007) Genotypic microbial community profiling: a critical technical review. Microb Ecol 54(2):276–289
131. Fischer SG, Lerman LS (1983) DNA fragments differing by single base-pair substitutions are separated in denaturing gradient gels: correspondence with melting theory. Proc Natl Acad Sci USA 80(6):1579–1583
132. Muyzer G, Smalla K (1998) "Application of denaturing gradient gel electrophoresis (DGGE) and temperature gradient gel electrophoresis (TGGE) in microbial ecology. Antonie van Leeuwenhoeck Int J Gen Mol Microbiol 73:127–141
133. Rosenbaum V, Riesner D (1987) Temperature-gradient gel electrophoresis. Thermodynamic analysis of nucleic acids and proteins in purified form and in cellular extracts. Biophys Chem 26(2–3):235–246
134. Orita M, Iwahana H, Kanazawa H, Hayashi K, Sekiya T (1989) Detection of polymorphisms of human DNA by gel electrophoresis as single-strand conformation polymorphisms. Proc Natl Acad Sci USA 86(8):2766–2770
135. Schwieger F, Tebbe CC (1998) A new approach to utilize PCR-single-strand-conformation polymorphism for 16S rRNA gene-based microbial community analysis. Appl Environ Microbiol 64(12):4870–4876
136. Su C, Lei L, Duan Y, Zhang K-Q, Yang J (2012) Culture-independent methods for studying environmental microorganisms: methods, application, and perspective. Appl Microbiol Biotechnol 93(3):993–1003
137. Wagner R (1994) The regulation of ribosomal RNA synthesis and bacterial cell growth. Arch Microbiol 161(2):100–109

138. Wintzingerode FV, Göbel UB, Stackebrandt E (2006) Determination of microbial diversity in environmental samples: pitfalls of PCR-based rRNA analysis. FEMS Microbiol Rev 21 (3):213–229
139. García-Martínez J, Acinas SG, Antón AI, Rodríguez-Valera F (1999) Use of the 16S-23S ribosomal genes spacer region in studies of prokaryotic diversity. J Microbiol Methods 36(1–2):55–64
140. Gürtler V, Stanisich A (1996) New approaches to typing and identification of bacteria using the 16S-23S rDNA spacer region. Microbiology 142 (Pt 1):3–16
141. Schmid M, Schmitz-Esser S, Jetten M, Wagner M (2001) 16S-23S rDNA intergenic spacer and 23S rDNA of anaerobic ammonium-oxidizing bacteria: implications for phylogeny and in situ detection. Environ Microbiol 3(7):450–459
142. Ziganshina EE, Bagmanova AR, Khilyas IV, Ziganshin AM (2014) Assessment of a biogas-generating microbial community in a pilot-scale anaerobic reactor. J Biosci Bioeng 117(6):730–736
143. Mills DK, Fitzgerald K, Litchfield CD, Gillevet PM (2003) A comparison of DNA profiling techniques for monitoring nutrient impact on microbial community composition during bioremediation of petroleum-contaminated soils. J Microbiol Methods 54(1):57–74
144. Stahl DA, Capman WC (1994) Applications of molecular genetics to the study of microbial communities. NATO ASI Ser 35:193–206
145. Muyzer G (2000) Genetic fingerprinting of microbial communities—present status and future perspectives. In: Bell CR, Brylinski M, Johnson-Green P (eds) Microbial biosystems: new frontiers. proceedings of the 8th international symposium on microbial ecology. Atlantic Canada Society for Microbial Ecology, Halifax, pp 503–572
146. Collins G, Woods A, McHugh S, Carton MW, O'Flaherty V (2003) Microbial community structure and methanogenic activity during start-up of psychrophilic anaerobic digesters treating synthetic industrial wastewaters. FEMS Microbiol Ecol 46(2):159–170
147. Padmasiri SI, Zhang J, Fitch M, Norddahl B, Morgenroth E, Raskin L (2007) Methanogenic population dynamics and performance of an anaerobic membrane bioreactor (AnMBR) treating swine manure under high shear conditions. Water Res 41(1):134–144
148. McHugh S, Collins G, O'Flaherty V (2006) Long-term, high-rate anaerobic biological treatment of whey wastewaters at psychrophilic temperatures. Bioresour Technol 97 (14):1669–1678
149. Scully C, Collins G, O'Flaherty V (2005) Assessment of anaerobic wastewater treatment failure using terminal restriction fragment length polymorphism analysis. J Appl Microbiol 99(6):1463–1471
150. Liesack W, Dunfield PF (2002) Biodiversity in soils: use of molecular methods for its characterization. In: Bitton G (ed) Encyclopedia of environmental microbiology. Wiley, New York, pp 4516–4522
151. Hugenholtz P, Goebel BM, Pace NR (1998) Impact of culture-independent studies on the emerging phylogenetic view of bacterial diversity. J Bacteriol 180(18):4765–4774
152. Schmalenberger A, Tebbe CC (2014) Profiling the diversity of microbial communities with single-strand conformation polymorphism (SSCP). Methods Mol Biol 1096:71–83
153. Smalla K, Oros-Sichler M, Milling A, Heuer H, Baumgarte S, Becker R, Neuber G, Kropf S, Ulrich A, Tebbe CC (2007) Bacterial diversity of soils assessed by DGGE, T-RFLP and SSCP fingerprints of PCR-amplified 16S rRNA gene fragments: do the different methods provide similar results? J Microbiol Methods 69(3):470–479
154. Chachkhiani M, Dabert P, Abzianidze T, Partskhaladze G, Tsiklauri L, Dudauri T, Godon JJ (2004) 16S rDNA characterisation of bacterial and archaeal communities during start-up of anaerobic thermophilic digestion of cattle manure. Bioresour Technol 93(3):227–232
155. Werner JJ, Knights D, Garcia ML, Scalfone NB, Smith S, Yarasheski K, Cummings TA, Beers AR, Knight R, Angenent LT (2011) Bacterial community structures are unique and resilient in full-scale bioenergy systems. Proc Natl Acad Sci USA 108(10):4158–4163
156. Hill TCJ, Walsh KA, Harris JA, Moffett BF (2003) Using ecological diversity measures with bacterial communities. FEMS Microbiol Ecol 43(1):1–11

157. Yu D, Kurola JM, Kymäläinen M, Sinkkonen A, Romantschuk M (2014) Biogas production and methanogenic archaeal community in mesophilic and thermophilic anaerobic co-digestion processes. J Environ Manage 143:54–60
158. De Vrieze J, Verstraete W, Boon N (2013) Repeated pulse feeding induces functional stability in anaerobic digestion. Microb Biotechnol 6(4):414–424
159. Franke-Whittle IH, Goberna M, Pfister V, Insam H (2009) Design and development of the ANAEROCHIP microarray for investigation of methanogenic communities. J Microbiol Methods 79(3):279–288
160. Franke-Whittle IH, Knapp BA, Fuchs J, Kaufmann R, Insam H (2009) Application of COMPOCHIP microarray to investigate the bacterial communities of different composts. Microb Ecol 57(3):510–521
161. Goberna M, Gadermaier M, Schoen MA, Sperl D, Franke-Whittle IH, Wett B, Insam H (2012) Fingerprinting the microbial communities in organic wastes using oligonucleotide microarrays and Real-Time PCR. In: Trasar-Cepeda C, Hernandez T, Garcia C, Rad C, Gonzales-Carcedo S (eds) Soil enzymology in the recycling of organic wastes and environmental restoration, Burgos, Spain. Springer, Berlin, pp 285–298
162. Mutz K-O, Heilkenbrinker A, Lönne M, Walter J-G, Stahl F (2013) Transcriptome analysis using next-generation sequencing. Curr Opin Biotechnol 24(1):22–30
163. Carvalhais LC, Dennis PG, Tyson GW, Schenk PM (2012) Application of metatranscriptomics to soil environments. J Microbiol Methods 91(2):246–251
164. Solomon KV, Haitjema CH, Thompson DA, O'Malley MA (2014) Extracting data from the muck: deriving biological insight from complex microbial communities and non-model organisms with next generation sequencing. Curr Opin Biotechnol 28:103–110
165. Quail MA, Smith M, Coupland P, Otto TD, Harris SR, Connor TR, Bertoni A, Swerdlow HP, Gu Y (2012) A tale of three next generation sequencing platforms: comparison of Ion Torrent, Pacific Biosciences and Illumina MiSeq sequencers. BMC Genom 13(1):341
166. Segata N, Boernigen D, Tickle TL, Morgan XC, Garrett WS, Huttenhower C (2013) Computational meta'omics for microbial community studies. Mol. Syst. Biol. 9(1):666
167. Koren S, Phillippy AM (2015) One chromosome, one contig: complete microbial genomes from long-read sequencing and assembly. Curr Opin Microbiol 23:110–120
168. Pinto AJ, Raskin L (2012) PCR biases distort bacterial and archaeal community structure in pyrosequencing datasets. PLoS ONE 7(8):e43093
169. Lebuhn M, Hanreich A, Klocke M, Schlüter A, Bauer C, Pérez CM (2014) Towards molecular biomarkers for biogas production from lignocellulose-rich substrates. Anaerobe 29:10–21
170. Zakrzewski M, Goesmann A, Jaenicke S, Jünemann S, Eikmeyer F, Szczepanowski R, Al-Soud WA, Sørensen S, Pühler A, Schlüter A (2012) Profiling of the metabolically active community from a production-scale biogas plant by means of high-throughput metatranscriptome sequencing. J Biotechnol 158(4):248–258
171. Wagner M, Assmus B, Hartmann A, Hutzler P, Amann R (1994) In situ analysis of microbial consortia in activated sludge using fluorescently labelled, rRNA-targeted oligonucleotide probes and confocal scanning laser microscopy. J Microsc 176(3):181–187
172. Ploem JS (1967) The use of a vertical illuminator with interchangeable dichroic mirrors for fluorescence microscopy with incident light. Z Wiss Mikrosk 68:129–142
173. Caldwell DE, Korber DR, Lawrence JR (1992) Confocal laser microscopy and digital image analysis in microbial ecology. Adv Microb Ecol 12:1–67
174. Lawrence JR, Korber DR, Hoyle BD, Costerton JW, Caldwell DE (1991) Optical sectioning of microbial biofilms. J Bacteriol 173(20):6558–6567
175. Taylor DL, Salmon ED (1989) Basic fluorescence microscopy. In: Wang Y-L, Taylor DL (eds) Fluorescence microscopy of living cells in culture. Academic Press, San Diego, pp 207–237

176. Behnam F, Vilcinskas A, Wagner M, Stoecker K (2012) A straightforward DOPE (double labeling of oligonucleotide probes)-FISH (fluorescence in situ hybridization) method for simultaneous multicolor detection of six microbial populations. Appl Environ Microbiol 78 (15):5138–5142
177. Amann R, Fuchs BM (2008) Single-cell identification in microbial communities by improved fluorescence in situ hybridization techniques. Nat Rev Microbiol 6(5):339–348
178. Amann R, Kühl M (1998) In situ methods for assessment of microorganisms and their activities. Curr Opin Microbiol 1:352–358
179. Raskin L, Poulsen LK, Noguera DR, Rittmann BE, Stahl DA (1994) Quantification of methanogenic groups in anaerobic biological reactors by oligonucleotide probe hybridization. Appl Environ Microbiol 60(4):1241–1248
180. Loy A, Horn M, Wagner M (2003) probeBase: an online resource for rRNA-targeted oligonucleotide probes. Nucleic Acids Res 31(1):514–516
181. Kepner RL, Pratt JR (1994) Use of fluorochromes for direct enumeration of total bacteria in environmental samples: past and present. Microbiol Rev 58(4):603–615
182. Sanz JL, Köchling T (2007) Molecular biology techniques used in wastewater treatment: An overview. Process Biochem 42(2):119–133
183. Diaper JP, Tither K, Edwards C (1992) Rapid assessment of bacterial viability by flow cytometry. Appl Microbiol Biotechnol 38(2)
184. Crosland-Taylor PJ (1953) A device for counting small particles suspended in a fluid through a Tube. Nature 171:37–38
185. Koch C, Harnisch F, Schröder U, Müller S (2014) Cytometric fingerprints: evaluation of new tools for analyzing microbial community dynamics. Front Microbiol 5:273
186. Wagner M, Ivleva NP, Haisch C, Niessner R, Horn H (2009) Combined use of confocal laser scanning microscopy (CLSM) and Raman microscopy (RM): investigations on EPS-Matrix. Water Res 43(1):63–76
187. Hoff K (1988) Rapid and simple method for double staining of bacteria with 4′,6-diamino-2-phenylindole and fluorescein isothiocyanate labelled antibodies. Appl Environ Microbiol 54(12):2949–2952
188. Lukinavičius G, Reymond L, D'Este E, Masharina A, Göttfert F, Ta H, Güther A, Fournier M, Rizzo S, Waldmann H, Blaukopf C, Sommer C, Gerlich DW, Arndt H-D, Hell SW, Johnsson K (2014) Fluorogenic probes for live-cell imaging of the cytoskeleton. Nat Methods 11(7):731–733
189. Grimm JB, Sung AJ, Legant WR, Hulamm P, Matlosz SM, Betzig E, Lavis LD (2013) Carbofluoresceins and carborhodamines as scaffolds for high-contrast fluorogenic probes. ACS Chem Biol 8(6):1303–1310
190. Hunter RC, Beveridge TJ (2005) Application of a pH-sensitive fluoroprobe (C-SNARF-4) for pH microenvironment analysis in Pseudomonas aeruginosa biofilms. Appl Environ Microbiol 71(5):2501–2510
191. Zebulun O (2013) Molecular and biological techniques used in landfill investigations: A mini-review. Biotechnol Mol Biol Rev 8(2):35–42
192. Lee N, Nielsen PH, Andreasen KH, Juretschko S, Nielsen JL, Schleifer KH, Wagner M (1999) Combination of fluorescent in situ hybridization and microautoradiography-a new tool for structure-function analyses in microbial ecology. Appl Environ Microbiol 65(3):1289–1297
193. Nielsen JL, Nielsen PH (2010) Combined microautoradiography and fluorescence in situ hybridization (MAR-FISH) for the identification of metabolically active microorganisms. In: Timmis KN (ed) Handbook of hydrocarbon and lipid microbiology. Springer, Berlin, pp 4093–4102
194. Collins G, Kavanagh S, McHugh S, Connaughton S, Kearney A, Rice O, Carrigg C, Scully C, Bhreathnach N, Mahony T, Madden P, Enright A-M, O'flaherty V (2006) Accessing the black box of microbial diversity and ecophysiology: recent advances through polyphasic experiments. J Environ Sci Health A Tox Hazard Subst Environ Eng 41(5):897–922

195. Okabe S, Kindaichi T, Ito T (2004) MAR-FISH—An ecophysiological approach to link phylogenetic affiliation and in situ metabolic activity of microorganisms at a single-cell resolution. Microbes Environ 19(2):83–98
196. Ariesyady HD, Ito T, Yoshiguchi K, Okabe S (2007) Phylogenetic and functional diversity of propionate-oxidizing bacteria in an anaerobic digester sludge. Appl Microbiol Biotechnol 75(3):673–683
197. Ariesyady HD, Ito T, Okabe S (2007) Functional bacterial and archaeal community structures of major trophic groups in a full-scale anaerobic sludge digester. Water Res 41 (7):1554–1568
198. Ito T, Yoshiguchi K, Ariesyady HD, Okabe S (2011) Identification of a novel acetate-utilizing bacterium belonging to Synergistes group 4 in anaerobic digester sludge. ISME J 5(12):1844–1856
199. Ito T, Yoshiguchi K, Ariesyady HD, Okabe S (2012) Identification and quantification of key microbial trophic groups of methanogenic glucose degradation in an anaerobic digester sludge. Bioresour Technol 123:599–607
200. Song H, Clarke WP, Blackall LL (2005) Concurrent microscopic observations and activity measurements of cellulose hydrolyzing and methanogenic populations during the batch anaerobic digestion of crystalline cellulose. Biotechnol Bioeng 91(3):369–378
201. Fernández N, Montalvo S, Fernández-Polanco F, Guerrero L, Cortés I, Borja R, Sánchez E, Travieso L (2007) Real evidence about zeolite as microorganisms immobilizer in anaerobic fluidized bed reactors. Process Biochem 42:721–728
202. Weiss S, Zankel A, Lebuhn M, Petrak S, Somitsch W, Guebitz GM (2011) Investigation of mircroorganisms colonising activated zeolites during anaerobic biogas production from grass silage. Bioresour Technol 102(6):4353–4359
203. Krakat N, Westphal A, Schmidt S, Scherer P (2010) Anaerobic digestion of renewable biomass: thermophilic temperature governs methanogen population dynamics. Appl Environ Microbiol 76(6):1842–1850
204. Cresson R, Dabert P, Bernet N (2009) Microbiology and performance of a methanogenic biofilm reactor during the start-up period. J Appl Microbiol 106(3):863–876
205. Karakashev D, Batstone DJ, Angelidaki I (2005) Influence of environmental conditions on methanogenic compositions in anaerobic biogas reactors. Appl Environ Microbiol 71 (1):331–338
206. Crocetti G, Murto M, Björnsson L (2006) An update and optimisation of oligonucleotide probes targeting methanogenic Archaea for use in fluorescence in situ hybridisation (FISH). J Microbiol Methods 65:194–201
207. Hoshino T, Yilmaz LS, Noguera DR, Daims H, Wagner M (2008) Quantification of target molecules needed to detect microorganisms by fluorescence in situ hybridization (FISH) and catalyzed reporter deposition-FISH. Appl Environ Microbiol 74(16):5068–5077
208. Stoecker K, Dorninger C, Daims H, Wagner M (2010) Double labeling of oligonucleotide probes for fluorescence in situ hybridization (DOPE-FISH) improves signal intensity and increases rRNA accessibility. Appl Environ Microbiol 76(3):922–926
209. Kumar RK, Chapple CC, Hunter N (1999) Improved double immunofluorescence for confocal laser scanning microscopy. J Histochem Cytochem 47:1213–1218
210. Ludwig W, Strunk O, Westram R, Richter L, Meier H, Yadhukumar Buchner A, Lai T, Steppi S, Jobb G, Förster WI, Brettske S, Gerber A, Ginhart W, Gross O, Grumann S, Hermann S, Jost R, König A, Liss T, Lüssmann R, May M, Nonhoff B, Reichel B, Strehlow R, Stamatakis A, Stuckmann N, Vilbig A, Lenke M, Ludwig T, Bode A, Schleifer K-H (2004) ARB: a software environment for sequence data. Nucleic Acids Res 32(4):1363–1371
211. Rotaru A-E, Shrestha PM, Liu F, Shrestha M, Shrestha D, Embree M, Zengler K, Wardman C, Nevin KP, Lovley DR (2014) A new model for electron flow during anaerobic digestion: direct interspecies electron transfer to Methanosaeta for the reduction of carbon dioxide to methane. Energy Environ Sci 7(1):408–415

212. Rotaru A-E, Shrestha PM, Liu F, Markovaite B, Chen S, Nevin K, Lovley D (2014) Direct interspecies electron transfer between Geobacter metallireducens and Methanosarcina barkeri. Appl Environ Microbiol 80(24):AEM.00895
213. Dubé CD, Guiot SR (2014) Direct interspecies electron transfer in granular sludge. In: International conference biogas science 2014, pp 85–86
214. Teng Z, Hua J, Wang C, Lu X (2014) Design and optimization principles of biogas reactors in large scale applications: research progress in lab—Immobilization. In: Shi F (ed) Reactor and process design in sustainable energy technology. Springer, Amsterdam, pp 125–127
215. Lalov IG, a. Krysteva M, Phelouzat JL (2001) Improvement of biogas production from vinasse via covalently immobilized methanogens. Bioresour Technol 79:83–85
216. Gong W, Liang H, Li W, Wang Z (2011) Selection and evaluation of biofilm carrier in anaerobic digestion treatment of cattle manure. Energy 36(5):3572–3578
217. Ahammad SZ, Davenport RJ, Read LF, Gomes J, Sreekrishnan TR, Dolfing J (2013) Rational immobilization of methanogens in high cell density bioreactors. RSC Adv 3:774
218. Bauer A, Theuretzbacher F, Rincón M, Enguidanos R (2014) Steam explosion pretreatment production of Alpine hay for enhancing biogas. In: Proceedings international conference of agricultural engineering, 2014, pp 4–7
219. Bharathiraja B, Sudharsanaa T, Bharghavi A, Sowmeya GS, Balaram G (2014) Insights on lignocellulosic pretreatments for biofuel production- SEM and reduction of lignin analysis, vol 6, no 9
220. Serrano-Lotina A, Daza L (2013) Highly stable and active catalyst for hydrogen production from biogas. J Power Sources 238:81–86
221. Knott G, Marchman H, Wall D, Lich B (2008) Serial section scanning electron microscopy of adult brain tissue using focused ion beam milling. J Neurosci 28(12):2959–2964
222. Ishitani T, Yaguchi T (1996) Cross-sectional sample preparation by focused ion beam: a review of ion-sample interaction. Microsc Res Tech 35(4):320–333

Anaerobic Fungi and Their Potential for Biogas Production

Veronika Dollhofer, Sabine Marie Podmirseg, Tony Martin Callaghan, Gareth Wyn Griffith and Kateřina Fliegerová

Abstract Plant biomass is the largest reservoir of environmentally friendly renewable energy on earth. However, the complex and recalcitrant structure of these lignocellulose-rich substrates is a severe limitation for biogas production. Microbial pro-ventricular anaerobic digestion of ruminants can serve as a model for improvement of converting lignocellulosic biomass into energy. Anaerobic fungi are key players in the digestive system of various animals, they produce a plethora of plant carbohydrate hydrolysing enzymes. Combined with the invasive growth of their rhizoid system their contribution to cell wall polysaccharide decomposition may greatly exceed that of bacteria. The cellulolytic arsenal of anaerobic fungi consists of both secreted enzymes, as well as extracellular multi-enzyme complexes called cellulosomes. These complexes are extremely active, can degrade both amorphous and crystalline cellulose and are probably the main reason of cellulolytic efficiency of anaerobic fungi. The synergistic use of mechanical and enzymatic degradation makes anaerobic fungi promising candidates to improve biogas

V. Dollhofer (✉)
Bavarian State Research Center for Agriculture, Central Department for Quality Assurance and Analytics, Micro- and Molecular Biology, Lange Point 6, 85354 Freising, Germany
e-mail: veronika.dollhofer@lfl.bayern.de

S.M. Podmirseg
Institut für Mikrobiologie, Universität Innsbruck, Technikerstraße 25d, 6020 Innsbruck, Austria
e-mail: Sabine.Podmirseg@uibk.ac.at

T.M. Callaghan · G.W. Griffith
Institute of Biological, Environmental and Rural Sciences, Aberystwyth University, Adeilad Cledwyn Penglais, Aberystwyth, Ceredigion SY23 3DD, Wales, UK
e-mail: toneocallaghan@hotmail.co.uk

G.W. Griffith
e-mail: gwg@aber.ac.uk

K. Fliegerová
Institute of Animal Physiology and Genetics, v.v.i, Academy of Sciences of the Czech Republic, 142 20, Prague 4, Czech Republic
e-mail: fliegerova@iapg.cas.cz

G.M. Guebitz et al. (eds.), *Biogas Science and Technology*,
Advances in Biochemical Engineering/Biotechnology 151,
DOI 10.1007/978-3-319-21993-6_2

production from recalcitrant biomass. This chapter presents an overview about their biology and their potential for implementation in the biogas process.

Keywords Anaerobic fungi · Neocallimastigomycota · Phylogeny · Cellulosomes · Biogas process improvement · Recalcitrant cellulosic substrates

Contents

1 Anaerobic Fungi: An Overview

Anaerobic fungi belonging to the phylum Neocallimastigomycota, are the most basal lineage of the kingdom Fungi. These fungi are principally known from the digestive tracts of larger mammalian herbivores, where they play an important role as primary colonisers of ingested forage [1, 2]. Recent studies indicate their appearance in herbivorous reptiles like the green iguana [2] and termites [3] also. Anaerobic fungi are characterised by several distinctive traits which stem from their obligately anaerobic physiology; mitochondria, cytochromes and other biochemical features of the oxidative phosphorylation pathway are absent. Energy generation occurs in hydrogenosomes where ATP is formed by malate decarboxylation to form acetate, CO_2, and H_2 [4]. The Neocallimastigales are fungi that do not require molecular oxygen for any of their physiological processes, and for which the presence of oxygen is toxic. This trait raises the question how anaerobic fungi defend themselves against the toxic effects of oxygen, for instance when colonizing freshly ingested forage or during dispersal between host animals. Respective insights are presented in the following section "life cycle". Additionally, their genomes are peculiar having the highest AT-content hitherto found (often exceeding 90 % in non-coding regions) and with a substantial expansion of important hydrolytic and cellulolytic gene families [5].

Anaerobic fungi are the only fungi which possess cellulosomes. These extraordinary features are presented in more detail in Sect. 2.1. The position of anaerobic fungi as a basal fungal lineage is reflected in the genome characteristics, which are also present in other early-branching fungal lineages and/or non-fungal Opisthokonts, but are absent in the later diverging Dikarya (Ascomycetes and Basidiomycetes) genomes [6]. Such phylogenetic determinants and unique taxonomy of anaerobic fungi are discussed in the following Sect. 1.1.

1.1 Classical and Pragmatic Taxonomy of Anaerobic Fungi

The atypical morphology and physiology of anaerobic fungi has caused some taxonomic uncertainty. After misleading classification as Protozoa [7], Phycomycetes [8] and Chytridiomycetes [9, 10] the anaerobic fungi were finally placed into the distinct phylum Neocallimastigomycota [11]. The phylum contains only one order (Neocallimastigales) and one family (Neocallimastigaceae) within which eight genera are currently described: The monocentric rhizoidal genera *Neocallimastix*, *Piromyces*, *Ontomyces* and *Buwchfawromyces*, the polycentric rhizoidal genera *Anaeromyces* and *Orpinomyces*, and the two bulbous genera, monocentric *Caecomyces* and polycentric *Cyllamyces*, respectively [12–14].

The genera are defined on the basis of thallus morphology, the formation of rhizoidal filaments or bulbous holdfasts within the substrate and their zoospore morphology. A distinction is made between monoflagellate and polyflagellate zoospores. The latter possessing 7–20 posterior flagella inserted in two rows. Formation of polyflagellate zoospores is a trait unique to *Orpinomyces* and *Neocallimastix* spp., not known from any other Opisthokonta, and these two genera form a distinct clade within the Neocallimastigomycota [15].

Differentiation by the shape of sporangia may additionally be possible, but can be misleading as it is varying depending on culture conditions. Currently about 20 species have been described [16]. Uncertainties created by difficulties in inter-lab comparisons and the loss of many viable type cultures, can only now be resolved by the use of DNA barcoding and the concerted effort to exchange cultures [17].

Culture-independent analysis of environmental nucleic acid sequences, provided evidence for much greater fungal diversity than previously suspected in the digestive tract of wild and domestic herbivores. Based on data from these more recent studies, it appears that twelve or more hitherto un-named genera may exist [2, 15, 18]. Several of these novel clades are now recognized from sequences of cultured fungi [15], while other clades still consist of environmental nucleic sequences (ENAS) only.

1.2 Life Cycle

The life cycle of anaerobic fungi alternates between a motile zoospore stage and a non-motile vegetative stage. The latter consists of a thallus associated to plant

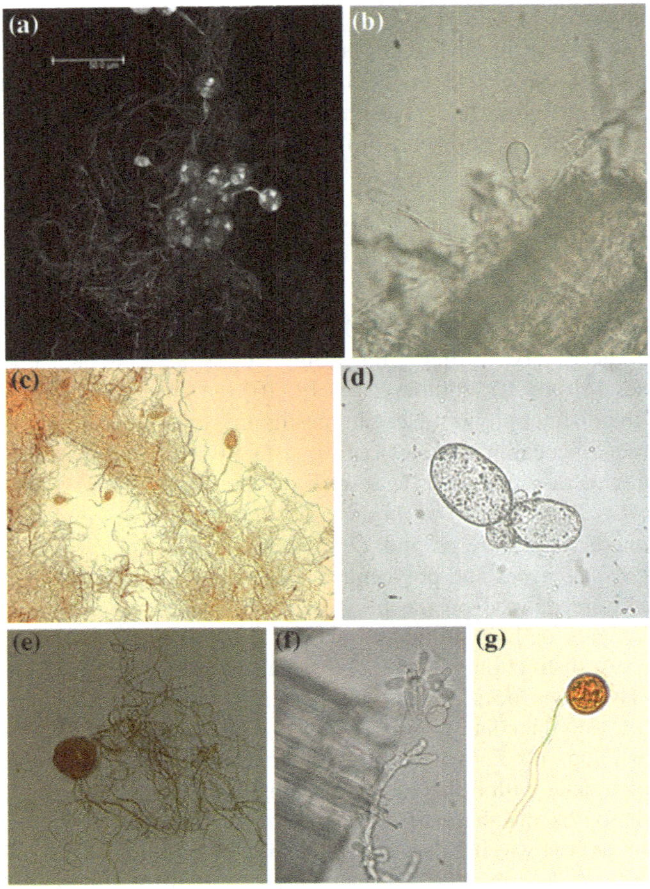

Fig. 1 Different culture morphologies of anaerobic fungi: **a** *Neocallimastix* sp. sporangia and rhizomycelium (CLSM: superimposed z-stacks (26.7 μm total depth) showing culture auto fluorescence (excitation at 561 nm and emission from 570 to 620 nm); **b** *Piromyces* sp. light microscopy of native preparation; **c** Rhizoid of *Anaeromyces mucronatus* with apical sporangia. Light microscopy of lugol-stained preparation (×200); **d** Bulbous species *Caecomyces communis*. Light microscopy of native preparation (×400); **e** *Neocallimastix frontalis* sporangium and rhizoid. Light microscopy (×400); **f** *Orpinomyces* sp. with sporangia and rhizoid. Light microscopy of native preparation; **g** Light microscopy of a biflagellated zoospore of *Piromyces* sp. (×1000)

material and fruiting bodies known as sporangia (Fig. 1) [13]. Flagellate zoospores (see Fig. 1g) released from mature sporangia actively swim towards freshly ingested plant tissues using chemotactic response to soluble sugars and/or phenolic acids [19]. After attachment to the feed particles, flagella are shed and a cyst is formed. The cyst then germinates to form the thallus. In all monocentric species (*Piromyces*, *Neocallimastix* and *Buwchfawromyces*), the nucleus remains in the enlarging cyst which forms the sporangium. In the polycentric species *Anaeromyces* and *Orpinomyces*, the nuclei migrate through the rhizoidal system to form multiple sporangia on a single thallus. The terms exogenous and endogenous germination (nuclei migrate into the thallus or not), that are widely used in describing chytrid development, are less clearly applicable to the bulbous anaerobic fungi which do not form rhizoids but do form multiple sporangia (i.e. *Cyllamyces*) [20].

The rhizoidal system penetrates the plant tissue by a combination of enzymatic activity and hydrostatic pressure using appressorium-like penetration structures [21, 22]. In the non-rhizoidal bulbous species (*Caecomyces, Cyllamyces*), the expanding holdfast formed within the substrate causes a splitting of the plant fibers [23–25]. Sporangium maturation and release of asexual zoospores can occur as quickly as eight hours after encystment [26, 27].The complete life cycle, is conducted within 24–32 h [25]. Propagules of the anaerobic fungi are known to survive up to and probably over a year in feces [28] and have also been found to be transferred to neonatal hosts through saliva [29]. Putative aero-tolerant survival structures have been observed only rarely [14, 30, 31] and many questions as to the formation of these structures and their occurrence in the various genera of anaerobic fungi remain to be answered.

1.3 Anaerobic Fungi and Their Interactions with Methanogens and Bacteria

Close association of anaerobic fungi with methanogens is well known [23, 32], with inter-species hydrogen transfer leading to both methane production and also more efficient re-generation of oxidized nucleotides (NAD^+, $NADP^+$). Syntrophic co-cultivation markedly increases fungal growth rate, with increased rates of cellulolysis and xylanolysis, consequently enhancing dry matter reduction [33]. However the anaerobic fungus—methanogen interaction is more complex than simple cross-feeding. Hydrogen transfer also influences fungal catabolic pathways and specific enzyme profiles, shifting fungal product formation away from more oxidized end products (lactate, ethanol) towards production of more reduced products (acetate, formate). Acetate, and in the rumen especially formate, are the preferred growth substrates for methanogens [32, 33]. This interaction is so pivotal, that some species of anaerobic fungi cannot be isolated as axenic cultures, but only in combination with the permanent archaeal symbiont [34].

Syntrophic interactions between acetogenic bacteria and methanogens are well known to occur in the biogas biocoenosis [35]. Since anaerobic fungi show improved growth in the presence of methanogens, the idea of augmenting biogas reactors with this microbial group seems a logical step.

Interactions of anaerobic fungi with bacteria can be of antagonistic and symbiotic nature as shown by Bernalier and coworkers [36], who tested the degradation efficiency in different culture combinations of three anaerobic fungi and two cellulolytic bacterial strains. In general both groups are competing for the same ecological niche, but the breaking up of plant tissue through fungal rhizoids may also enhance the overall efficiency of cellulolytic bacteria [36]. This improved degradation was also confirmed when testing the contribution of different microbial groups (fungi, bacteria, protozoa) on orchard grass decomposition [37]. Presence of protozoa was, however attributed with lower degradation efficiency and inhibition of both, bacteria and fungi.

Most of these studies are based on in vitro co-cultures, that may not completely reflect conditions of whole rumen or biogas reactor consortia and still more research is needed in this field.

2 Anaerobic Fungi and Their Potential for Biogas Production

Under oxygen-free conditions organic matter is decomposed by a complex of microorganisms which are so far divided into three functional groups: hydrolysing and fermenting bacteria, obligate hydrogen-producing acetogenic bacteria, and methanogenic archaea. Only little is known on the role and the potential of anaerobic fungi for biogas production. Great potential lies in biogas production from lignocellulosic wastes but, slow and inefficient degradation processes, the formation of toxic intermediates and the necessity for long incubation times are only a few examples of the problems encountered [38, 39]. A promising strategy is the use of microorganisms, which are able to successfully perform such complicated degradation processes in their natural environment [40, 41]. Herbivores as biogas reactors evolved the need for fungal symbionts for this purpose and over millions of years natural selection has created a highly specialised and niche specific community of anaerobic fungi.

The following paragraphs will give an overview about useful features of anaerobic fungi and will present the actual knowledge about anaerobic fungi and biogas production.

2.1 Lignocelluloytic Enzymes of Anaerobic Fungi and Their Potential Use

Lignin-embedded cellulose and hemicellulose [42] represent a physical barrier against microbial and enzymatic attack. Known as the primary digesters of plant biomass in the rumen anaerobic fungi [37] have the ability to open up the plant tissue through rhizoidal growth and produce a cocktail of enzymes to degrade and separate the different compounds of lignocellulosic biomass, while lignin itself remains anaerobically indigestible. Some of these enzymes are secreted freely but most of them are bound to a multi-enzyme complex the so called cellulosome.

Genome sequencing of *Orpinomyces* strain C1A revealed a broader enzyme range compared to aerobic fungi with a repertoire of 357 glycosyl hydrolases, 92 carbohydrate esterases and 24 pectate lyases [5]. Horizontal gene transfer from bacteria is suggested as one of the main reasons why anaerobic fungi have evolved such robust and impressive cellulolytic and hemicellulolytic capability.

A group of enzymes often termed cellulases synergistically hydrolyze β-1, 4 glucosidic bonds in cellulose through three discrete enzymatic activities involving three different types of enzymes. Endoglucanases (EC 3.2.1.4) cut within amorphous regions of cellulose strands, releasing oligosaccharides and creating new free chain ends for the enzymatic attack by exoglucanases (EC 3.2.1.176; EC 3.2.1.91). Since the latter liberate cellobiose disaccharides from either reducing (EC 3.2.1.176) or non-reducing (EC 3.2.1.91) ends, they are also termed cellobiohydrolases. In a cellulosomal complex extracted from a *Neocallimastix frontalis* culture, enzymes from glycosyl hydrolase family 5 (GH5) operated by the endo- and enzymes from GH6 and GH48 by the exo-mechanism [43]. The residual cellobiose is then hydrolyzed to glucose by β-glucosidases (EC 3.2.1.21) [40, 44]. Auxiliary enzymes like the recently discovered lytic polysaccharide mono-oxygenases (LPMO) (family AA9) have been reported to enhance or complete the utilization of cellulose in many fungal species [45]. In contrast to the hydrolyzing enzymes they cleave glucosidic bonds with a copper dependent oxidation mechanism and are able to attack crystalline regions of cellulose [46]. But it seems that basal fungal groups including the anaerobic fungi lack those enzymes [45].

All three major cellulase types have been reported for the Neocallimastigomycota ([5, 47, 48, 49, 50] and many more) confirming the potential of anaerobic fungi as a reservoir for highly efficient cellulases. The fact that glucose is the main product of anaerobic fungal cellulose degradation is an advantage for biotechnological applications. Cellobiose is not accumulated and therefore cannot act as end-product inhibitor for cellulose hydrolysis, as known for *Trichoderma reesei* or many bacterial species. Thus costly addition of β-glucosidase becomes unnecessary [51].

Due to the heterogeneous structure of hemicelluloses, several enzymes are needed for their catabolism. Until now anaerobic fungi have been reported to provide all enzymes needed to degrade the major hemicelluloses constituents of the plant cell wall, namely β-glucans, mannans and xylans. And in some cases xylanase activity was even higher than cellulase activity [52]. In contrast to aerobic higher fungi (Dikarya), anaerobic fungi lack the enzymatic machinery to catabolise lignin. The enzymatic reaction to cleave the aromatic ring requires oxygen and can therefore not take place in an anaerobic environment [53]. But it was shown that a *Neocallimastix* sp. could mediate the loss of up to 34 % of plant biomass associated lignin, however this loss probably due to physical alteration or chemical modification of the lignin rather than enzymatic catabolism [54]. Additional feruloyl (EC 3.1.1.73) esterases are produced which cleave the bond between hemicelluloses and lignin and by separating these two compounds, making cellulose and hemicellulose more easily accessible for further degradation [55].

2.1.1 Anaerobic Fungal Cellulosomes

As mentioned above, most of the cellulolytic and hemicellulolytic enzymes are part of a multi-enzyme complex known as the cellulosome. Cellulosomes were first identified in the bacterial family *Clostridiaceae* [56] and the anaerobic fungi are the only eukaryotic representatives showing this feature. The fungal cellulosome is structurally and phylogenetically similar to that found in bacteria and is thought to have arisen through a horizontal gene transfer event [57]. Up to now cellulosomes have been described for species of *Piromyces* [58, 59], *Orpinomyces* [48], and *Neocallimastix* [52, 60]. Anaerobic fungi invade plant tissues with their rhizoid and it is assumed that in addition to the secretion of soluble enzymes, they form cellulosomes anchored to the cell walls of rhizoid tips [55]. Unfortunately the molecular structure of the anaerobic fungal cellulosome is still unclear and miscellaneous theories exist (see [61] for a schematic overview). In anaerobic bacteria a non-catalytic protein, the 'scaffolding protein', is anchored to the cell wall and contains several repeating domains, the cohesins. This structure forms the backbone to which the enzymatic subunits assemble by non-catalytic domains, the dockerins. Additionally the scaffolding connects to the substrate, in this case the (hemi) cellulose molecules, via a cellulose-binding domain [62].

Compared to the enzymes of anaerobic bacteria, which contain only one species-specific dockerin domain, the fungal enzymes contain one to three copies of dockerin domains which show an interspecies specificity. It is believed that the amount of dockerin regulates the affinity of the enzymes towards the scaffolding molecule [63]. Recently it was reported that the anaerobic fungal cellulosome contains a scaffolding backbone as well, raising the suggestion that the catalytic components also interact with it via dockerin domains [43]. Other studies have shown that some types of docking domains attach to several individual proteins, concluding that there might be various different scaffolding proteins in anaerobic fungal cellulosomes [64]. Additionally it could be shown that a double-dockerin domain and a β-glucosidase enzymatic subunit from glycosyl hydrolase family 3 (GH 3), both belonging to one fungal species, could bind to each other [58, 61]. This leads to the third theory that dockerins mediate the binding of different secreted enzymes to each other, forming the cellulosome without scaffolding as structural molecule. Despite the detailed structure remaining unsolved, cellulosomes permit the anaerobic fungi to use their cellulolytic enzymes in a synergistic and more efficient way, unequalled by individually secreted enzymes [61]. It also provides protection against proteases from the surrounding environment in the form of a serine protease inhibitor named celpin [65].

2.1.2 Substrates Utilized by Anaerobic Fungi

In addition to municipal solid waste (MSW) and animal wastes, lignocellulose-rich materials potentially useful for biogas production are by-products of various industrial processes, including agriculture, forestry, pulp-, paper- and food

Table 1 Examples for lignocellulosic residues degraded by anaerobic fungi

Lignocellulosic residue	Lignin content % [66]	Organism	Reference
Wheat straw	16-21	*Neocallimastix frontalis*	[67]
Coastal Bermuda grass	6.4	*Piromyces* MC-1, *Orpinomyces* PC-1-3, *Neocallimastix* MC-2	[49]
Sugar cane bagasse	19-24	*Piromyces* strain E2	[68]
Hard wood	18-25	*Neocallimastix* sp.	[69]
Rice straw	18	*Piromyces* M014, *Orpinomyces* GSRI-001, *Neocallimastix* T010	[3]

production [51, 66]. However, the recalcitrance and variability of these materials leads to low gas yields in biogas fermentations, thus making their exploitation uneconomical. Since anaerobic fungi are efficient physical and enzymatic degraders of lignocellulose-rich substrates (see Table 1), they have the potential to make the biogas production from these lignocellulose-rich materials more efficient and profitable.

2.1.3 Production of Recombinant Enzymes

One strategy to overcome the bottleneck of enzymatic hydrolysis of lignocellulose in the biogas production process is the development and use of recombinant potent polysaccharide-degrading enzymes. Such a strategy could involve the transfer of the cellulolytic genes of efficient degraders (e.g. anaerobic fungi) into other well-established enzyme production hosts or biofuel producers (e.g. yeast) or alternatively the modification of the genetic capability of the anaerobic fungi themselves. Improving the efficiency of known enzymes and the creation of optimized enzyme mixtures, along with the identification of new and more active enzymes has been the focus of some studies [70]. Efforts to produce recombinant fibrolytic enzymes from anaerobic fungi have focused on expressing a range of carbohydrate-active enzymes into a number of aerobic fungal expression hosts. But catalytic activity of anaerobic fungal xylanases, cellulases, β-glucosidases, or cellobiohydrolases in the tested aerobic strains (*Saccharomyces cerevisiae*, *Hypocrea jecorina, Pichia pastoris* and *P. methanolica*) was low or else the recombinant proteins were not catalytically active [71–74]. Genetic modification of *S. cerevisiae* integrating a xylose isomerase from anaerobic fungi allowing the yeast to metabolize monosaccharide xylose was more successful. Conversion of xylose into xylulose using the isomerases of *Piromyces* and/or *Orpinomyces* species [75–77] represents at this time the most promising technique for improving the industrial production of ethanol [78] and several patents have been filed so far [79]. In addition to the incorporation of single enzymes, the creation of artificial cellulosomes and xylanosomes, to profit from the synergy between certain enzymes is on the rise. For example Doi and colleagues built a cellulosome from *Clostridium thermocellum* enzymes which show synergistic activity against cellulose [66]. Mingardon et al. designed mini-cellulosomes combining free fungal endoglucanase of glycosyl hydrolase family 6 from *Neocallimastix patriciarum* with bacterial cellulosomal endoglucanase of glycosyl hydrolase family 9 from *Clostridium cellulolyticum*, achieving superior cellulose activity, compared to complexes assembled only with bacterial enzymes [80]. But even if recombinant anaerobic fungal enzymes could be produced and implemented in biotechnological processes, the physical degradation abilities of anaerobic fungi would still remain unused.

2.2 Anaerobic Fungi in the Biogas Production Process

A commonly encountered issue during anaerobic digestion is limited degradability of plant biomass, 40–60 % of organic carbon remains unused [81]. This problem is due to the physical structure and the recalcitrant chemical nature of these polymers. In detail, lignin remains indigestive under anaerobic conditions and shields cellulose and hemicellulose from enzymatic degradation. Thus, technologies that can improve anaerobic degradation of lignocellulosic biomass are needed. Partial disruption of plant tissues, can be achieved by mechanical [82], thermal [83, 84], chemical [85], oxidative [86] or ultrasonic [87, 88] pre-treatment.

However, in the rumen the natural biogas system these techniques are not available. There bacteria, archaea, protozoa and anaerobic fungi account for the key players in plant tissue degradation. Some important parameters of anaerobic digestion in biogas fermenters resemble conditions of the fermentation processes found in the rumen, namely a strong negative redox potential, a nearly neutral pH and a temperature between 37 ± 2 °C. Microbial pre-treatment or the implementations of rumen microorganisms into the biogas process seem to be possible strategies to deal with recalcitrant substrates.

Improvement of anaerobic biomass hydrolysis through the addition of specific microorganisms has been experimentally tested in several studies for bacteria. Miah and co-workers [89] described a 210 % increase in biogas production during thermophilic digestion (65 °C) of sewage sludge caused by the protease activity of a *Geobacillus* sp. strain. Similarly, Bagi and colleagues [90] applied mesophilic *Enterobacter cloacae* and thermophilic *Caldicellulosyruptor saccharolyticus* strains during anaerobic digestion of waste water sludge, pig manure and dried plant biomass of artichoke, and achieved a remarkable increase of biogas production (160 %). This increase was explained by the enhanced H_2 level as both tested strains are excellent hydrogen-producing bacteria, and *C. saccharolyticus* has moreover cellulolytic activity. Also introduction of an aerobic pre-treatment step for plant residues through e.g. white and brown rot fungi or the potent cellulose degrading *Trichoderma viride* has shown promising results on improving the subsequent anaerobic digestibility in biogas reactors [91, 92].

In contrast, the direct introduction of anaerobic fungi into these bioreactors would eliminate the requirement of an aerobic pre-digestion. With respect to the presented intention, of course only mesophilic conditions are eligible. In recent years, several studies have dealt with the application of anaerobic fungi to improve anaerobic digestion of cellulosic material [3]. In more detail, the digestive tract of animals fed with very specific, fibre-rich diets have been chosen for the isolation of potent anaerobic fungal strains, that could be best suited for a technical implementation [34]. The possibility of *Anaeromyces* and *Piromyces* strains to integrate into biogas-producing anaerobic sludge bacterial communities, to improve degradation of substrate polysaccharides and consequently to influence methane production has already been tested in laboratory conditions. Promising results were obtained during the bioaugmentation of swine manure fed biogas reactors with different strains of anaerobic fungi. Amendment with fungal biomass led to 4–22 % higher gas yields and up to 2.5 % higher methane concentration [81, 93]. A recent study showed that bioaugmentation with anaerobic fungi did not increase the overall methane yield, but that it speeds up initial gas production and thus may help to reduce retention time [94]. In most cases, however, it was not possible to preserve fungal activity and the fungal beneficial effect on hydrolysis seems to decline after about ten days of incubation. The factors permitting fungal growth in habitats other than the digestive tract of their hosts still require thorough research and it is unclear if full-scale application of these microorganisms will become feasible.

3 Anaerobic Fungi: Methodological State of the Art

3.1 Detection Techniques for Anaerobic Fungi

The monitoring of anaerobic fungi sampled from the digestive tract or feces of herbivores requires accurate and reliable detection techniques, and the same methods are also applicable to axenic cultures and industrial fermentations [95]. Here we summarize the range of approaches that have been used so far, or which may be of relevance to detect and quantify the activity of anaerobic fungi.

Microscopy is still the most straightforward method for a general determination of growth status and initial phylogenetic classification of fungal biomass. However it requires a certain level of skill and experience to assign identity and mistakes can be made even with the help of identification keys as found in e.g. Ho and Barr [96] and Orpin [97]. Classification into rhizoidal or bulbous genera is relatively easy, for a more exact attribution of anaerobic fungi to the monocentric or polycentric group, the DNA binding fluorescent dyes DAPI (4’,6-diamidino-2-phenylindole) or stains of the Hoechst-group (bisbenzimides) must be employed. A microscopic approach reaches its limit when differentiation between e.g. *Piromyces* and *Neocallimastix*, or *Orpinomyces* and *Anaeromyces* is needed and often no zoospore release can be witnessed to check for monoflagellate or polyflagellate zoospores. Another drawback, especially in microscopy of environmental samples that contain plant debris, is the clear differentiation of fungal- and plant biomass. During fluorescence microscopy, autofluorescence of plant material over a wide wavelength range clearly impedes distinct identification of fungal structures. Staining with Calcofluor white [98] or the more recently proposed stains Solophenyl Flavine 7GFE 500 and Pontamine Fast Scarlet 4B [99] will help to highlight chitinous structures of the fungal biomass, such as cell walls, septa and bud scars, but the affinity of these dyes for cellulose and other sugar polymers can be problematic. Specific staining protocols can be performed to circumvent this issue. One possibility is the staining with lactofuchsin as described in Leis et al. [34], an approach originally used to bring out plant root fungi, e.g. arbuscular myccorhizas.

Measurement of fungal abundance with culture-dependent techniques i.e. thallus forming units (TFU) is generally performed through the most probable number (MPN) method [29, 100] and by using the roll-tube method as described by Joblin [101]. A work that can be tedious and also requires certain expertise. The roll-tube approach is further well suited to obtain pure fungal cultures during the isolation procedure.

An indirect way to determine fungal biomass/growth is through their gas production that can be monitored by the use of a pressure transducer and then correlated to the amount of biomass [102].

Anaerobic fungi produce a wide range of potent enzymes, e.g. cellulase, endoglucanase, xylanase or amylase amongst others, that help to degrade plant material [93, 103, 104]. Thus enzyme activity can be used as indirect parameter for fungal activity. For instance Fliegerová and co-workers could, based on these parameters, demonstrate the improved hydrolytic activity of biogas reactors after fungal amendment, but also detected the relatively fast decrease of this enzyme activity over time [93].

Another very promising approach that has yet to be tested for anaerobic fungi is the raising of enzyme-specific antibodies. Li and coworkers [103] were able to produce specific antibodies for the catalytic domain of xylanases found in *Orpinomyces* and *Neocallimastix*. By fluorescence-labelling of these antibodies that could maybe also be raised for other fungi specific structures, an elegant detection technique could be established.

Culture independent, molecular techniques and DNA-based approaches have revolutionized microbial ecology over the last two decades and helped to confirm the monophyly of the Neocallimastigomycota. The most commonly used target genes, that allow not only for anaerobic fungi detection and community analysis but also quantification through qPCR are the small ribosomal subunit (18S rRNA gene) and the internal spacer (ITS) region [15, 32, 95, 105–109]. However, both gene regions also bear certain drawbacks that should be considered and are discussed in [13]. To summarize these drawbacks, the sequence of the 18S rRNA gene is too conserved within the Neocallimastigomycota phylum to allow for a clear differentiation of closely related taxa [110], and the ITS region, despite its prevalent utilization in fungal phylogeny [111], does not allow for direct sequencing of PCR products and exhibits high variability for this microbial group that might impair sequence alignments. The 28S rRNA gene however seems to be best suited for detection and phylogenetic assignment of anaerobic fungi and should be considered as the best target gene thus far utilized. A recent study even suggests to combine all three DNA regions (18S, 28S and ITS) for a more accurate representation of fungal diversity in environmental samples [112], indicating that each chosen DNA region leads to a different result. Quantification of anaerobic fungi through qPCR gives a good insight into fungal abundance but is difficult to correlate with culture dependent enumeration results (TFU) or the actual biomass due to varying ratios of the DNA/biomass content within the Neocallimastigomycota members and depending on specific growth phase of each culture.

3.2 Cultivation Techniques and Cryopreservation

This chapter has highlighted the potential of this unusual group of fungi to address a range of problems associated with the degradation of lignocellulose-rich waste materials. The fact that these fungi are obligate anaerobes is an important component of their biotechnological potential, since scale-up issues are less problematic with anaerobic fermentation. However, the associated difficulty in the culturing and maintenance of obligate anaerobic fungi does impede the exchange of materials between scientists, and could cause problems in future biotechnological deployment of these fungi. First there is a need for an international culture collection, with moves underway to exchange cryogenically stored cultures between interested parties. This will avoid the loss of cultures that has beset past research—we note with sadness that most of the type cultures that define the ca. 20 species are no longer extant. However, the growth in the routine use of DNA barcoding will facilitate the process of reliable identification of these fungi both in pure culture and from environmental samples.

Storage in liquid nitrogen appears to provide the only means for long term storage of anaerobic fungi cultures and it is strongly advised to store such cryovials in several locations. Storage at −80 °C is possible but there is progressive loss of viability of cultures over periods of more than a few months. Given the fragility of pure cultures, there is a need to elucidate the mechanism whereby these fungi form aerotolerant structures. It is clear that all the anaerobic fungi must be able to do this in order to disperse between hosts and furthermore it is clear that they are very efficient in dispersal. The ability to generate such aerotolerant structures from axenic cultures would be extremely useful for long-term preservation of cultures and important in the context of this chapter for the inoculation of industrial fermentations with desired cultures or culture mixtures. Fliegerová et al. [93] has already demonstrated that biogas fermentation can be enhanced by addition of anaerobic fungi, as have Puniya et al. in their use of 'direct fed' microbials for the enhancement of the rumen fermentation [113]. However, they used actively growing cultures, a process difficult to scale up. The ability to add aerotolerant structures to such fermentations would be most advantageous.

4 Conclusions

One of the major research goals in biogas science is to find an efficient tool to circumnavigate the bottleneck possessed by hydrolysis of lignocellulose-rich residues. Besides several physical, mechanical chemical or microbial pretreatment techniques, the use of anaerobic lignocellulolytic fungi should be beneficial and even more cost-efficient. The rumen of herbivores can be seen as a natural resource for potent biomass degraders. Especially anaerobic fungi, known to act as primary digesters, could be good candidates.

They produce a superior set of hemi/cellulolytic enzymes which they excrete separately or combined in cellulosomes. Additionally they are able to attack the plant material mechanically by their rhizoidal growth and open up the tissue for further digestion by bacteria. These two features are of capital interest to the biogas industry.

Until now several attempts have been made to produce recombinant anaerobic fungal enzymes for biotechnological application and even artificial cellulosomes have been built. Production in yeast has been the most profitable way, but still more research has to be done to provide recombinant enzymes in an industrial scale. Experiments to use anaerobic fungi directly in the biogas production process showed positive effects on gas production, but enzymatic activity and fungal growth decreased quickly under these conditions. Maybe anaerobic fungi cannot be implemented into conventional biogas reactors, but an individual anaerobic fungal pre-hydrolysis stage might be a possible solution facing this problem.

To summarize, anaerobic fungi have the potential to make biogas production much more efficient and the utilization of lignocellulose-rich substrates more viable. But for use in the industrial scale a greater understanding of the underlying ecology of these fungi and there cohorts is needed.

Acknowledgments TMC is grateful for funding from the Aberystwyth Postgraduate Research Studentship. VD is grateful for funding of the project BE/14/22 from the Bavarian State Ministry of Food, Agriculture and Forestry and the Bavarian State Ministry of Economics.

References

1. Akin DE, Borneman WS (1990) Role of rumen fungi in fiber degradation. J Dairy Sci 73 (10):3023–3032. doi:10.3168/jds.S0022-0302(90)78989-8
2. Liggenstoffer AS, Youssef NH, Couger MB, Elshahed MS (2010) Phylogenetic diversity and community structure of anaerobic gut fungi (phylum Neocallimastigomycota) in ruminant and non-ruminant herbivores. ISME J 4(10):1225–1235. doi:10.1038/ismej.2010.49
3. Lee SM, Guan LL, Eun JS, Kim CH, Lee SJ, Kim ET, Lee SS (2015) The effect of anaerobic fungal inoculation on the fermentation characteristics of rice straw silages. J Appl Microbiol 118(3):565–573. doi:10.1111/jam.12724
4. Van der Giezen M (2009) Hydrogenosomes and mitosomes: conservation and evolution of functions. J Eukaryot Microbiol 56(3):221–231. doi:10.1111/j.1550-7408.2009.00407.x

5. Youssef NH, Couger MB, Struchtemeyer CG, Liggenstoffer AS, Prade RA, Najar FZ, Atiyeh HK, Wilkins MR, Elshahed MS (2013) The genome of the anaerobic fungus *Orpinomyces* sp. strain C1A reveals the unique evolutionary history of a remarkable plant biomass degrader. Appl Environ Microbiol 79(15):4620–4634. doi:10.1128/AEM.00821-13
6. Liu YJ, Hodson MC, Hall BD (2006) Loss of the flagellum happened only once in the fungal lineage: phylogenetic structure of kingdom Fungi inferred from RNA polymerase II subunit genes. BMC Evol Biol 6:74. doi:10.1186/1471-2148-6-74
7. Liebetanz E (1910) Die parasitischen Protozoen des Wiederkäuermagens. Arch Protistenk 19:19–90
8. Orpin CG (1977) The occurrence of chitin in the cell walls of the rumen organisms *Neocallimastix frontalis*, *Piromonas communis* and *Sphaeromonas communis*. J Gen Microbiol 99(1):215–218. doi:10.1099/00221287-99-1-215
9. Barr DJS (1980) An outline for the reclassification of the Chytridiales, and for a new order, the Spizellomycetales. Can J Bot 58(22):2380–2394. doi: 10.1139/b80-276
10. Barr DJS (1988) How modern systematics relates to the rumen fungi. Bio Syst 21(3–4):351–356. doi:10.1016/0303-2647(88)90032-9
11. Hibbett DS, Binder M, Bischoff JF, Blackwell M, Cannon PF, Eriksson OE, Huhndorf S, James T, Kirk PM, Lücking R, Thorsten Lumbsch H, Lutzoni F, Matheny PB, McLaughlin DJ, Powell MJ, Redhead S, Schoch CL, Spatafora JW, Stalpers JA, Vilgalys R, Aime MC, Aptroot A, Bauer R, Begerow D, Benny GL, Castlebury LA, Crous PW, Dai YC, Gams W, Geiser DM, Griffith GW, Gueidan C, Hawksworth DL, Hestmark G, Hosaka K, Humber RA, Hyde KD, Ironside JE, Kõljalg U, Kurtzman CP, Larsson KH, Lichtwardt R, Longcore J, Miadlikowska J, Miller A, Moncalvo JM, Mozley-Standridge S, Oberwinkler F, Parmasto E, Reeb V, Rogers JD, Roux C, Ryvarden L, Sampaio JP, Schüßler A, Sugiyama J, Thorn RG, Tibell L, Untereiner WA, Walker C, Wang Z, Weir A, Weiss M, White MM, Winka K, Yao YJ, Zhang N (2007) A higher-level phylogenetic classification of the Fungi. Mycol Res 111(5):509–547. doi:10.1016/j.mycres.2007.03.004
12. Dagar SS et al. 2015 A new anaerobic fungus (Oontomyces anksri gen. nov., sp. nov.) from the digestive tract of the Indian camel (Camelus dromedarius). Fungal Biology
13. Gruninger RJ, Puniya AK, Callaghan TM, Edwards JE, Youssef N, Dagar SS, Fliegerová K, Griffith GW, Forster R, Tsang A, McAllister T, Elshahed MS (2014) Anaerobic fungi (phylum Neocallimastigomycota): advances in understanding their taxonomy, life cycle, ecology, role and biotechnological potential. FEMS Microbiol Ecol 90(1):1–17. doi:10.1111/1574-6941.12383
14. Callaghan TM, Podmirseg SM, Hohlweck D, Edwards JE, Puniya AK, Dagar SS, Griffith GW (2015) *Buwchfawromyces eastonii* gen. nov., sp. nov.: a new anaerobic fungus (Neocallimastigomycota) isolated from buffalo faeces. MycoKeys 9:11–28. doi:10.3897/mycokeys.9.9032
15. Koetschan C, Kittelmann S, Lu J, Al-Halbouni D, Jarvis GN, Muller T, Wolf M, Janssen PH (2014) Internal transcribed spacer 1 secondary structure analysis reveals a common core throughout the anaerobic fungi (Neocallimastigomycota). PLoS ONE 9(3):e91928. doi:10.1371/journal.pone.0091928
16. Fliegerová K, Mrázek J, Voigt K (2012) Gut fungi: classification, evolution, life style and application. In: Fungi: types, environmental impact and role in disease, pp 3–19
17. Griffith GW, Baker S, Fliegerová K, Liggenstoffer AS, van der Giezen M, Voigt K, Beakes G (2010) Anaerobic fungi: Neocallimastigomycota. IMA Fungus 1(2):181–185
18. Fliegerová K, Mrázek J, Hoffmann K, Zábranská J, Voigt K (2010) Diversity of anaerobic fungi within cow manure determined by ITS1 analysis. Folia Microbiol 55(4):319–325. doi:10.1007/s12223-010-0049-y
19. Wubah DA, Kim DSH (1996) Chemoattraction of anaerobic ruminal fungi zoospores to selected phenolic acids. Microbiol Res 151(3):257–262. doi:10.1016/S0944-5013(96)80022-X

20. Ozkose E, Thomas BJ, Davies DR, Griffith GW, Theodorou MK (2001) *Cyllamyces aberensis* gen. nov. sp. nov., a new anaerobic gut fungus with branched sporangiophores isolated from cattle. Can J Bot 79(6):666–673. doi:10.1139/b01-047
21. Ho YW, Bauchop T (1991) Morphology of three polycentric rumen fungi and description of a procedure for the induction of zoosporogenesis and release of zoospores in cultures. J Gen Microbiol 137(1):213–217. doi:10.1099/00221287-137-1-213
22. Ho YW, Abdullah N, Jalaludin S (1988) Penetrating structures of anaerobic rumen fungi in cattle and swamp buffalo. J Gen Microbiol 134(1):177–181. doi:10.1099/00221287-134-1-177
23. Orpin CG, Joblin K (1997) The rumen anaerobic fungi. In: The rumen microbial ecosystem. Springer, Berlin, pp 140–195
24. Orpin CG (1977) The rumen flagellate *Piromonas communis*: its life-history and invasion of plant material in the rumen. J Gen Microbiol 99(1):107–117. doi:10.1099/00221287-99-1-107
25. Lowe SE, Griffith GW, Milne A, Theodorou MK, Trinci APJ (1987) The life cycle and growth kinetics of an anaerobic rumen fungus. J Gen Microbiol 133(7):1815–1827. doi:10.1099/00221287-133-7-1815
26. France J, Theodorou MK, Davies DR (1990) Use of zoospore concentrations and life cycle parameters in determining the population of anaerobic fungi in the rumen ecosystem. J Theor Biol 147(3):413–422. doi:10.1016/S0022-5193(05)80496-5
27. Theodorou MK, Davies DR, Jordan MGC, Trinci APJ, Orpin CG (1993) Comparison of anaerobic fungi in faeces and rumen digesta of newly born and adult ruminants. Mycol Res 97(10):1245–1252. doi:10.1016/S0953-7562(09)81293-8
28. McGranaghan P, Davies JC, Griffith GW, Davies DR, Theodorou MK (1999) The survival of anaerobic fungi in cattle faeces. FEMS Microbiol Ecol 29(3):293–300. doi:10.1016/S0168-6496(99)00024-0
29. Davies DR, Theodorou MK, Lawrence MIG, Trinci APJ (1993) Distribution of anaerobic fungi in the digestive tract of cattle and their survival in faeces. J Gen Microbiol 139(Pt 6):1395–1400. doi:10.1099/00221287-139-6-1395
30. Brookman JL, Ozkose E, Rogers S, Trinci APJ, Theodorou MK (2000) Identification of spores in the polycentric anaerobic gut fungi which enhance their ability to survive. FEMS Microbiol Ecol 31(3):261–267. doi:10.1111/j.1574-6941.2000.tb00692.x
31. Wubah DA, Fuller MS, Akin DE (1991) Resistant body formation in Neocallimastix sp., an anaerobic fungus from the rumen of a cow. Mycologia 83(1):40–47
32. Cheng YF, Edwards JE, Allison GG, Zhu WY, Theodorou MK (2009) Diversity and activity of enriched ruminal cultures of anaerobic fungi and methanogens grown together on lignocellulose in consecutive batch culture. Bioresour Technol 100(20):4821–4828. doi:10.1016/j.biortech.2009.04.031
33. Nakashimada Y, Srinivasan K, Murakami M, Nishio N (2000) Direct conversion of cellulose to methane by anaerobic fungus *Neocallimastix frontalis* and defined methanogens. Biotechnol Lett 22(3):223–227. doi:10.1023/A:1005666428494
34. Leis S, Dresch P, Peintner U, Fliegerovà K, Sandbichler AM, Insam H, Podmirseg SM (2014) Finding a robust strain for biomethanation: anaerobic fungi (Neocallimastigomycota) from the Alpine ibex (*Capra ibex*) and their associated methanogens. Anaerobe 29:34–43. doi:10.1016/j.anaerobe.2013.12.002
35. Stams AJ (1994) Metabolic interactions between anaerobic bacteria in methanogenic environments. Antonie Van Leeuwenhoek 66(1–3):271–294. doi:10.1007/BF00871644
36. Bernalier A, Fonty G, Bonnemoy F, Gouet P (1992) Degradation and fermentation of cellulose by the rumen anaerobic fungi in axenic cultures or in association with cellulolytic bacteria. Curr Microbiol 25(3):143–148. doi:10.1007/BF01571022
37. Lee SS, Ha JK, Cheng K-J (2000) Relative contributions of bacteria, protozoa, and fungi to in vitro degradation of orchard grass cell walls and their interactions. Appl Environ Microbiol 66(9):3807–3813. doi:10.1128/AEM.66.9.3807-3813.2000
38. Čater M, Zorec M, Logar RM (2014) Methods for improving anaerobic lignocellulosic substrates degradation for enhanced biogas production. Springer Science Reviews. doi:10.1007/s40362-014-0019-x

39. Kabir MM, Forgács G, Sárvári Horváth I (2015) Biogas from Lignocellulosic Materials. In: Karimi K (ed) Lignocellulose-Based Bioproducts, vol 1. Biofuel and Biorefinery Technologies. Springer International Publishing, pp 207–251. doi:10.1007/978-3-319-14033-9_6
40. Lynd LR, Weimer PJ, van Zyl WH, Pretorius IS (2002) Microbial cellulose utilization: fundamentals and biotechnology. Microbiol Mol Biol Rev 66 (3):506–577. doi:10.1128/MMBR.66.3.506-577.2002
41. Wen B, Yuan X, Li QX, Liu J, Ren J, Wang X, Cui Z (2015) Comparison and evaluation of concurrent saccharification and anaerobic digestion of Napier grass after pretreatment by three microbial consortia. Bioresour Technol 175:102–111. doi:10.1016/j.biortech.2014.10.043
42. Cullen D, Kersten P (1992) Fungal enzymes for lignocellulose degradation. In: Applied molecular genetics of filamentous fungi. Chapman and Hall, London, pp 100–131
43. Wang HC, Chen YC, Hseu RS (2014) Purification and characterization of a cellulolytic multienzyme complex produced by *Neocallimastix patriciarum* J11. Biochem Biophys Res Commun 451(2):190–195. doi:10.1016/j.bbrc.2014.07.088
44. Juturu V, Wu JC (2014) Microbial cellulases: engineering, production and applications. Renew Sustain Energy Rev 33:188–203. doi:10.1016/j.rser.2014.01.077
45. Morgenstern I, Powlowski J, Tsang A (2014) Fungal cellulose degradation by oxidative enzymes: from dysfunctional GH61 family to powerful lytic polysaccharide monooxygenase family. Briefings Funct Genomics 13(6):471–481. doi:10.1093/bfgp/elu032
46. Hemsworth GR, Henrissat B, Davies GJ, Walton PH (2014) Discovery and characterization of a new family of lytic polysaccharide monooxygenases. Nat Chem Biol 10(2):122–126. doi:10.1038/nchembio.1417
47. Harhangi HR, Steenbakkers PJ, Akhmanova A, Jetten MS, van der Drift C, Op den Camp HJ (2002) A highly expressed family 1 beta-glucosidase with transglycosylation capacity from the anaerobic fungus *Piromyces* sp. E2. Biochim Biophys Acta 1574(3):293–303. doi:10.1016/S0167-4781(01)00380-3
48. Ljungdahl LG (2008) The Cellulase/Hemicellulase System of the Anaerobic Fungus *Orpinomyces* PC-2 and Aspects of Its Applied Use. Ann N Y Acad Sci 1125 (1):308–321. doi:10.1196/annals.1419.030
49. Borneman WS, Hartley RD, Morrison WH, Akin DE, Ljungdahl LG (1990) Feruloyl and p-coumaroyl esterase from anaerobic fungi in relation to plant cell wall degradation. Appl Microbiol Biotechnol 33(3):345–351. doi:10.1007/BF00164534
50. Borneman WS, Akin DE, Ljungdahl LG (1989) Fermentation products and plant cell wall-degrading enzymes produced by monocentric and polycentric anaerobic ruminal fungi. Appl Environ Microbiol 55(5):1066–1073
51. Dashtban M, Schraft H, Qin W (2009) Fungal bioconversion of lignocellulosic residues; opportunities & perspectives. Int J Bio Sci 5(6):578. doi:10.7150/ijbs.5.578
52. Aylward JH, Gobius KS, Xue GP, Simpson GD, Dalrymple BP (1999) The *Neocallimastix patriciarum* cellulase, CelD, contains three almost identical catalytic domains with high specific activities on Avicel. Enzyme Microb Technol 24(8):609–614. doi:10.1016/S0141-0229(98)00167-7
53. Berg B (2014) Decomposer Organisms. In: Berg B, Mc Claugherty C (eds)Plant Litter-Decomposition, Humus Formation, Carbon Sequestration, 3rd edn. Springer, Heidelberg, pp 35–52. doi:10.1007/978-3-642-38821-7_3
54. McSweeney CS, Dulieu A, Katayama Y, Lowry JB (1994) Solubilization of lignin by the ruminal anaerobic fungus *Neocallimastix patriciarum*. Appl Environ Microbiol 60(8):2985–2989
55. Nagpal R, Puniya AK, Griffith GW, Goel G, Puniya M, Sehgal J, Paul, Singh K (2009) Anaerobic rumen fungi: potential and applications. In: Khachatourians GG, Arora DK,

Rajendran TP, Srivastava AK (eds) Agriculturally important microorganisms, vol 1. Academic World International, New York, pp 375–393
56. Felix CR, Ljungdahl LG (1993) The cellulosome: the exocellular organelle of Clostridium. Ann Rev Microbiol 47(1):791–819. doi:10.1146/annurev.mi.47.100193.004043
57. Garcia-Vallvé S, Romeu A, Palau J (2000) Horizontal gene transfer of glycosyl hydrolases of the rumen fungi. Mol Biol Evol 17(3):352–361
58. Nagy T, Tunnicliffe RB, Higgins LD, Walters C, Gilbert HJ, Williamson MP (2007) Characterization of a double dockerin from the cellulosome of the anaerobic fungus *Piromyces equi*. J Mol Biol 373(3):612–622. doi:10.1016/j.jmb.2007.08.007
59. Steenbakkers PJ, Freelove A, Van Cranenbroek B, Sweegers BM, Harhangi HR, Vogels GD, Hazlewood GP, Gilbert HJ, Op den Camp HJ (2002) The major component of the cellulosomes of anaerobic fungi from the genus *Piromyces* is a family 48 glycoside hydrolase. DNA Seq J DNA Seq Mapp 13(6):313–320. doi:10.1080/1042517021000024191
60. Zhou L, Xue GP, Orpin CG, Black GW, Gilbert HJ, Hazlewood GP (1994) Intronless celB from the anaerobic fungus *Neocallimastix patriciarum* encodes a modular family A endoglucanase. Biochem J 297(Pt 2):359–364
61. Haitjema CH, Solomon KV, Henske JK, Theodorou MK, O'Malley MA (2014) Anaerobic gut fungi: advances in isolation, culture, and cellulolytic enzyme discovery for biofuel production. Biotechnol Bioeng 111(8):1471–1482. doi:10.1002/bit.25264
62. Shoham Y, Lamed R, Bayer EA (1999) The cellulosome concept as an efficient microbial strategy for the degradation of insoluble polysaccharides. Trends Microbiol 7(7):275–281. doi:10.1016/S0966-842X(99)01533-4
63. Fontes CM, Gilbert HJ (2010) Cellulosomes: highly efficient nanomachines designed to deconstruct plant cell wall complex carbohydrates. Annu Rev Biochem 79:655–681. doi:10.1146/annurev-biochem-091208-085603
64. Steenbakkers PJ, Li XL, Ximenes EA, Arts JG, Chen H, Ljungdahl LG, Op Den Camp HJ (2001) Noncatalytic docking domains of cellulosomes of anaerobic fungi. J Bacteriol 183 (18):5325–5333. doi:10.1128/JB.183.18.5325-5333.2001
65. Steenbakkers PJ, Irving JA, Harhangi HR, Swinkels WJ, Akhmanova A, Dijkerman R, Jetten MS, van der Drift C, Whisstock JC, Op den Camp HJ (2008) A serpin in the cellulosome of the anaerobic fungus *Piromyces* sp. strain E2. Mycol Res 112(Pt 8):999–1006. doi:10.1016/j.mycres.2008.01.021
66. Sánchez C (2009) Lignocellulosic residues: biodegradation and bioconversion by fungi. Biotechnol Adv 27(2):185–194. doi:10.1016/j.biotechadv.2008.11.001
67. Li X, Calza RE (1991) Fractionation of cellulases from the ruminal fungus *Neocallimastix frontalis* EB188. Appl Environ Microbiol 57(11):3331–3336
68. Teunissen MJ, de Kort GV, Op den Camp HJ, Vogels GD (1993) Production of cellulolytic and xylanolytic enzymes during growth of anaerobic fungi from ruminant and nonruminant herbivores on different substrates. Appl Biochem Biotechnol 39–40:177–189. doi:10.1007/BF02918988
69. Joblin KN, Naylor GE (1989) Fermentation of woods by rumen anaerobic fungi. FEMS Microbiol Lett 65(1):119–122. doi:10.1111/j.1574-6968.1989.tb03608.x
70. Biswas R, Persad A, Bisaria VS (2014) Production of Cellulolytic Enzymes. In: Bisaria VS, Kondo A (eds) Bioprocessing of renewable resources to commodity bioproducts, John Wiley & Sons, pp 105–132. doi:10.1002/9781118845394
71. Li XL, Skory CD, Ximenes EA, Jordan DB, Dien BS, Hughes SR, Cotta MA (2007) Expression of an AT-rich xylanase gene from the anaerobic fungus *Orpinomyces* sp. strain PC-2 in and secretion of the heterologous enzyme by *Hypocrea jecorina*. Appl Microbiol Biotechnol 74(6):1264–1275. doi:10.1007/s00253-006-0787-6
72. Tsai CT, Huang C-T (2008) Overexpression of the *Neocallimastix frontalis* xylanase gene in the methylotrophic yeasts *Pichia pastoris* and *Pichia methanolica*. Enzyme Microb Technol 42(6):459–465. doi:10.1016/j.enzmictec.2008.01.018

73. Van Wyk N, Den Haan R, Van Zyl W (2010) Heterologous production of NpCel6A from *Neocallimastix patriciarum* in *Saccharomyces cerevisiae*. Enzyme Microb Technol 46 (5):378–383. doi:10.1016/j.enzmictec.2009.11.005
74. O'Malley MA, Theodorou MK, Kaiser CA (2012) Evaluating expression and catalytic activity of anaerobic fungal fibrolytic enzymes native topiromyces sp E2 in *Saccharomyces cerevisiae*. Environ Prog Sustain Energy 31(1):37–46. doi:10.1002/ep.10614
75. Kuyper M, Hartog MM, Toirkens MJ, Almering MJ, Winkler AA, Dijken JP, Pronk JT (2005) Metabolic engineering of a xylose-isomerase-expressing *Saccharomyces cerevisiae* strain for rapid anaerobic xylose fermentation. FEMS Yeast Res 5(4–5):399–409. doi:10.1016/j.femsyr.2004.09.010
76. Van Maris AJ, Winkler AA, Kuyper M, De Laat WT, Van Dijken JP, Pronk JT (2007) Development of efficient xylose fermentation in *Saccharomyces cerevisiae*: xylose isomerase as a key component. In: Olsson L (ed) Biofuels. Springer, Berlin pp 179–204. doi:10.1007/10_2007_057
77. Madhavan A, Tamalampudi S, Ushida K, Kanai D, Katahira S, Srivastava A, Fukuda H, Bisaria VS, Kondo A (2009) Xylose isomerase from polycentric fungus *Orpinomyces*: gene sequencing, cloning, and expression in *Saccharomyces cerevisiae* for bioconversion of xylose to ethanol. Appl Microbiol Biotechnol 82(6):1067–1078. doi:10.1007/s00253-008-1794-6
78. Bellissimi E, Van Dijken JP, Pronk JT, Van Maris AJ (2009) Effects of acetic acid on the kinetics of xylose fermentation by an engineered, xylose-isomerase-based *Saccharomyces cerevisiae* strain. FEMS Yeast Res 9(3):358–364. doi:10.1111/j.1567-1364.2009.00487.x
79. Teunissen AWRH, De Bont JAM (2011) Xylose isomerase genes and their use in fermentation of pentose sugars. US 13/142,124, 29 Dec 2011
80. Mingardon F, Chanal A, López-Contreras AM, Dray C, Bayer EA, Fierobe H-P (2007) Incorporation of fungal cellulases in bacterial minicellulosomes yields viable, synergistically acting cellulolytic complexes. Appl Environ Microbiol 73(12):3822–3832. doi:10.1128/AEM.00398-07
81. Procházka J, Mrázek J, Štrosová L, Fliegerová K, Zábranská J, Dohányos M (2012) Enhanced biogas yield from energy crops with rumen anaerobic fungi. Eng Life Sci 12 (3):343–351. doi:10.1002/elsc.201100076
82. Nah IW, Kang YW, Hwang K-Y, Song W-K (2000) Mechanical pretreatment of waste activated sludge for anaerobic digestion process. Water Res 34(8):2362–2368. doi:10.1016/S0043-1354(99)00361-9
83. Climent M, Ferrer I, Baeza MdM, Artola A, Vázquez F, Font X (2007) Effects of thermal and mechanical pretreatments of secondary sludge on biogas production under thermophilic conditions. Chem Eng J 133(1):335–342. doi:10.1016/j.cej.2007.02.020
84. Bougrier C, Delgenès J, Carrère H (2007) Impacts of thermal pre-treatments on the semi-continuous anaerobic digestion of waste activated sludge. Biochem Eng J 34(1):20–27. doi:10.1016/j.bej.2006.11.013
85. Tanaka S, Kamiyama K (2002) Thermochemical pretreatment in the anaerobic digestion of waste activated sludge. Water Sci Technol J Int Assoc Water Pollut Res 46(10):173–179. doi:10.1016/S0273-1223(97)00169-8
86. Goel R, Tokutomi T, Yasui H (2003) Anaerobic digestion of excess activated sludge with ozone pretreatment. Water Sci Technol J Int Assoc Water Pollut Res 47(12):207–214
87. Neis U, Nickel K, Tiehm A (2000) Enhancement of anaerobic sludge digestion by ultrasonic disintegration. Water Sci Technol 42(9):73–80
88. Appels L, Dewil R, Baeyens J, Degrève J (2008) Ultrasonically enhanced anaerobic digestion of waste activated sludge. Int J Sustain Eng 1(2):94–104. doi:10.1080/19397030802243319
89. Miah MS, Tada C, Sawayama S (2004) Enhancement of biogas production from sewage sludge with the addition of *Geobacillus* sp. strain AT1 culture. 日本水処理生物学会誌 40 (3):97–104. doi:10.2521/jswtb.40.97

90. Bagi Z, Ács N, Bálint B, Horváth L, Dobó K, Perei KR, Rákhely G, Kovács KL (2007) Biotechnological intensification of biogas production. Appl Microbiol Biotechnol 76(2):473–482. doi:10.1007/s00253-007-1009-6
91. Ghosh A, Bhattacharyya B (1999) Biomethanation of white rotted and brown rotted rice straw. Bioprocess Eng 20(4):297–302. doi:10.1007/s004490050594
92. Wagner AO, Schwarzenauer T, Illmer P (2013) Improvement of methane generation capacity by aerobic pre-treatment of organic waste with a cellulolytic *Trichoderma viride* culture. J Environ Manage 129:357–360. doi:10.1016/j.jenvman.2013.07.030
93. Fliegerová K, Procházka J, Mrázek J, Novotná Z, Štrosová L, Dohányos M (2012) Biogas and rumen fungi. In: Litonjua R, Cvetkovski I (eds) Biogas: production, consumption and applications, Nova Science Pub Inc, pp 161–180
94. Nkemka VN, Gilroyed B, Yanke J, Gruninger R, Vedres D, McAllister T, Hao X (2015) Bioaugmentation with an anaerobic fungus in a two-stage process for biohydrogen and biogas production using corn silage and cattail. Biores Technol 185:79–88. doi:10.1016/j.biortech.2015.02.100
95. Lockhart RJ, Van Dyke MI, Beadle IR, Humphreys P, McCarthy AJ (2006) Molecular biological detection of anaerobic gut fungi (Neocallimastigales) from landfill sites. Appl Environ Microbiol 72(8):5659–5661. doi:10.1128/AEM.01057-06
96. Ho Y, Barr D (1995) Classification of anaerobic gut fungi from herbivores with emphasis on rumen fungi from Malaysia. Mycologia 87(5):655–677. doi:10.2307/3760810
97. Orpin CG (1994) Anaerobic fungi: taxonomy, biology and distribution in nature. In: Mountfort DC, Orpin CG (eds) Anaerobic fungi: biology, ecology and function, CRC Press, pp 1–46
98. Darken MA (1961) Applications of fluorescent brighteners in biological techniques. Science 133(3465):1704–1705. doi:10.1126/science.133.3465.1704
99. Hoch HC, Galvani CD, Szarowski DH, Turner JN (2005) Two new fluorescent dyes applicable for visualization of fungal cell walls. Mycologia 97(3):580–588. doi:10.3852/mycologia.97.3.580
100. Theodorou MK, Gill M, King-Spooner C, Beever DE (1990) Enumeration of anaerobic chytridiomycetes as thallus-forming units: novel method for quantification of fibrolytic fungal populations from the digestive tract ecosystem. Appl Environ Microbiol 56(4):1073–1078
101. Joblin KN (1981) Isolation, enumeration, and maintenance of rumen anaerobic fungi in roll tubes. Appl Environ Microbiol 42(6):1119–1122
102. Theodorou MK, Davies DR, Nielsen BB, Lawrence MI, Trinci APJ (1995) Determination of growth of anaerobic fungi on soluble and cellulosic substrates using a pressure transducer. Microbiology 141(3):671–678. doi:10.1099/13500872-141-3-671
103. Li XL, Chen H, Ljungdahl LG (1997) Monocentric and polycentric anaerobic fungi produce structurally related cellulases and xylanases. Appl Environ Microbiol 63(2):628–635
104. Novotná Z, Procházka J, Simůnek J, Fliegerová K (2010) Xylanases of anaerobic fungus *Anaeromyces mucronatus*. Folia Microbiol 55(4):363–367. doi:10.1007/s12223-010-0059-9
105. Denman SE, McSweeney CS (2006) Development of a real-time PCR assay for monitoring anaerobic fungal and cellulolytic bacterial populations within the rumen. FEMS Microbiol Ecol 58(3):572–582. doi:10.1111/j.1574-6941.2006.00190.x
106. Edwards JE, Kingston-Smith AH, Jimenez HR, Huws SA, Skot KP, Griffith GW, McEwan NR, Theodorou MK (2008) Dynamics of initial colonization of nonconserved perennial ryegrass by anaerobic fungi in the bovine rumen. FEMS Microbiol Ecol 66(3):537–545. doi:10.1111/j.1574-6941.2008.00563.x
107. Fliegerová K, Mrázek J, Voigt K (2006) Differentiation of anaerobic polycentric fungi by rDNA PCR-RFLP. Folia Microbiol 51(4):273–277. doi:10.1007/BF02931811
108. Lwin KO, Hayakawa M, Ban-Tokuda T, Matsui H (2011) Real-time PCR assays for monitoring anaerobic fungal biomass and population size in the rumen. Curr Microbiol 62 (4):1147–1151. doi:10.1007/s00284-010-9843-7

109. Marano AV, Gleason FH, Barlocher F, Pires-Zottarelli CL, Lilje O, Schmidt SK, Rasconi S, Kagami M, Barrera MD, Sime-Ngando T, Boussiba S, de Souza JI, Edwards JE (2012) Quantitative methods for the analysis of zoosporic fungi. J Microbiol Methods 89(1):22–32. doi:10.1016/j.mimet.2012.02.003
110. Dagar SS, Kumar S, Mudgil P, Singh R, Puniya AK (2011) D1/D2 domain of large-subunit ribosomal DNA for differentiation of *Orpinomyces* spp. Appl Environ Microbiol 77 (18):6722–6725. doi:10.1128/AEM.05441-11
111. Schoch CL, Seifert KA, Huhndorf S, Robert V, Spouge JL, Levesque CA, Chen W, Fungal Barcoding C, Fungal Barcoding Consortium Author L (2012) Nuclear ribosomal internal transcribed spacer (ITS) region as a universal DNA barcode marker for Fungi. Proc Natl Acad Sci USA 109(16):6241–6246. doi:10.1073/pnas.1117018109
112. Tan H, Cao L (2014) Fungal diversity in sheep (*Ovis aries*) and cattle (*Bos taurus*) feces assessed by comparison of 18S, 28S and ITS ribosomal regions. Ann Microbiol 64(3):1423–1427. doi:10.1007/s13213-013-0787-6
113. Sehgal JP, Jit D, Puniya AK, Singh K (2008) Influence of anaerobic fungal administration on growth, rumen fermentation and nutrient digestion in female buffalo calves. J Anim Feed Sci 17:510–518

Hygiene and Sanitation in Biogas Plants

Bianca Fröschle, Monika Heiermann, Michael Lebuhn, Ute Messelhäusser and Matthias Plöchl

Abstract The increasing number of agricultural biogas plants and higher amounts of digestate spread on agricultural land arouse a considerable interest in the hygiene situation of digested products. This chapter reviews the current knowledge on sanitation during anaerobic digestion and the hygienic status of digestate concerning a multitude of pathogens potentially compromising the health of humans, animals and plants. Physical, chemical and biological parameters influencing the efficiency of sanitation in anaerobic digestion are considered. The degree of germ reduction depends particularly on the resistance of the pathogen of concern, the processing conditions, the feedstock composition and the diligence of the operation management. Most scientific studies facing sanitation in biogas plants have provided data ascertaining reduction of pathogens by the biogas process. Some pathogens, however, are able to persist virtually unaffected due to the ability to build resistant permanent forms. As compared to the feedstock, the sanitary status of the digestate is thus improved or in the worst case, the sanitary quality remains almost unchanged. According to this, the spreading of digestate on agricultural area in accordance to current rules and best practice recommendations is considered to impose no additional risk for the health of humans, animals and plants.

B. Fröschle (✉) · M. Lebuhn
Department for Quality Assurance and Analytics, Bavarian State Research Center for Agriculture, Lange Point 6, 85354 Freising, Germany
e-mail: froeschle.bianca@gmx.de

M. Lebuhn
e-mail: michael.lebuhn@lfl.bayern.de

M. Heiermann
Department Technology Assessment and Substance Cycles, Leibniz-Institute for Agricultural Engineering Potsdam-Bornim, Max-Eyth-Allee 100, 14469 Potsdam, Germany

U. Messelhäusser
Bavarian Health and Food Safety Authority, Veterinärstraße 2, 85764 Oberschleißheim, Germany

M. Plöchl
BioenergieBeratungBornim GmbH, Max-Eyth-Allee 101, 14469 Potsdam, Germany

G.M. Guebitz et al. (eds.), *Biogas Science and Technology*,
Advances in Biochemical Engineering/Biotechnology 151,
DOI 10.1007/978-3-319-21993-6_3

Contents

1 Introduction

Anaerobic digestion has gained increasing importance over the last decades. Formerly applied for the treatment of sewage, a trend toward the digestion of livestock husbandry residues, crop residues and lastly agricultural renewable resources like "energy crops" for biogas production has evolved [1]. Even biological wastes and animal by-products are commonly disposed by applying them as feedstock for biogas plants.

Besides the production of methane and heat a further valuable product of anaerobic digestion is its effluent. Digestate is usually spread on arable land as fertilizer. It provides comparable or slightly improved nitrogen recovery characteristics compared to undigested liquid animal slurry by increased ammonium nitrogen share on total nitrogen. The ammonium nitrogen fraction approximately corresponds to the plant available nitrogen fraction [2].

Increasing distribution and capacity of agricultural biogas production lead to large amounts of digestate spread on agricultural fields. In 2012, Möller and Müller [2] calculated a volume of 65.5 million cubic meters of digestate annually in Germany. As digestate and animal slurry can contain pathogens potentially eliciting human, animal and plant diseases, spreading entails the risk to transmit them to farmlands. In contrast to traditional manure management, large biogas plants operated by two or more owners or a company often obtain manure and crops from different farmers. Digestates may be spread on different farmers' fields or due to regional overflow transported over considerable distances. This practice can cause broader distribution of potentially contaminated material.

In this context, the growth of the biogas sector has aroused considerable interest in the presence and behavior of pathogens in anaerobic digestion processes and its effluents, but has also led to denunciations of agricultural biogas plants being cause of sanitary problems, which has not been scientifically ascertained. In Germany, assertions were claimed by the public and in media that toxin-producing clostridia (particularly *Clostridium botulinum*, a spore-forming, gram-positive bacterium causing botulism by strong biological neurotoxins), pathogenic *Salmonella* spp. and *Escherichia coli* pathovars proliferate in anaerobic digestion and are distributed by spreading digestate of agricultural biogas plants.

However, according to the state of science, the sanitary status of digestate is improved compared to the feedstock—even in manure digestion lines. Basically, the use of biogas technology does improve the operational and environmental sanitary situation. As the chapter of this book will show, meanwhile, a multitude of scientifically sound studies ascertains the reduction of pathogens by the process steps of ensiling, manure storage and particularly the biogas process. The degree of the reduction depends primarily on the resistance of the pathogen and the processing conditions. In general, the efficiency of sanitation, i.e. the degree of germ reduction, in the biogas process is influenced by physical, chemical and biological parameters. These are the topic of Chap. 2. As sanitation depends strongly on the pathogen species of concern, Chaps. 3 and 4 intend to review the current knowledge on qualitative and quantitative data of indicator organisms and diverse pathogens of plant, animal and human diseases in biogas processes.

2 Parameters Influencing the Efficiency of Sanitation in Anaerobic Digestion

2.1 Temperature

Particularly in thermophilic anaerobic digestion (≥55 °C), the hygienization potential proved to be high. Research on this topic was mainly done in the context of waste water treatment and digestion of sewage sludge (e.g. [3]), but also in connection with anaerobic digestion in biogas plants. A multitude of studies analyzing the existence and behavior of pathogens in biogas processes is proving that the sanitary effect of anaerobic digestion on pathogens increases with process temperature [4–12]. Most pathogens can endure better at mesophilic conditions (approx. 35–42 °C) than at thermophilic conditions (approx. 45–60 °C), but in both cases they were reduced.

Pathogens are usually adapted to the body temperature of their host which normally is between 30 and 40 °C for mammals and slightly higher for birds (35–45 °C) [13]. Optimal growth of human and animal pathogens therefore takes place in this temperature range. Plant pathogens usually can survive in a wide range of temperatures, spanning seasonal minimum and maximum values. Some can grow on plant parts in epiphyte growth stages and survive e.g. on seeds or soil between the seasons [14]. Temperatures between 16 and 24 °C, that are below the optimal growth conditions (25–30 °C), often influence the expression of virulence factors and cause pathogenicity in plants, e.g. at 18–20 °C in *Pseudomonas syringae* that optimally grows at 28 °C [15, 16].

In general, higher temperatures compared to the optimum growth temperature rapidly cause decrease in the growth rate and inactivation or death, whereas lower temperatures can rather be endured [17, 18]. In contrast, permanent forms (e.g. spores) are normally able to overcome extreme temperatures.

As agricultural biogas plants were predominantly operated under mesophilic conditions in a temperature range between 37 and 42 °C [19], the effect of temperature alone should not be the crucial factor for inactivation of human and veterinary pathogens but may affect phytopathogenic organisms. However, thermophilic temperatures provide unfavorable conditions to most pathogens.

Driven by increasing temperature, rising ammonia concentrations further impair pathogenic organisms in digesters. The equilibrium between ammonium and ammonia shifts with increasing temperature strongly in favor of ammonia that is toxic to bacteria [20]. Thereby, thermophilic anaerobic digestion of substrates rich in protein (e.g. grass silage) is connected with fast enrichment of ammonia (often accompanied by process disturbances, see also Sect. 2.4) [21, 22].

2.2 pH Value

The majority of pathogens favor a pH value close to the neutral range. This is in agreement with optimum conditions for anaerobic digestion that lie between pH values 6.5 and 8.5 [19, 23–26]. In contrast, fungi prefer slightly acidic conditions [27]. The influence of the pH value on the decay of pathogens in the biogas process depends on the divergence of the pH tolerance of the considered pathogen and the pH value prevailing in the digester of concern.

Besides, changes in pH value affect the availability of other substances that probably support inactivation of pathogens. An increasing pH value influences the equilibrium between ammonium and ammonia strongly in favor of ammonia [20, 21, 28]. High ammonia concentrations are toxic to most bacteria. In contrast, a study revealed a more toxic effect on rotaviruses at pH 6.9 (with mainly ammonium present) than at pH 9, which was related to an increasing sulfide concentration in raw wastewaters [20]. Similarly, Ottoson et al. [29] found an improved reduction of bacteriophages applied as a model organism for enteric viruses at a low ammonia concentration.

2.3 Retention Time and Exposition

The term retention time has to be differentiated into the minimum guaranteed retention time (MGRT) and the hydraulic retention time (HRT). The HRT is a statistical and calculative value. It describes the quotient of the active volume of the biogas plant (or the considered compartment) and the volume of added substrate per day. The MGRT is referred to as the time which a particle spends in the conditions of the digester. In stirred tank reactors, which are most common in agricultural biogas plants, the MGRT is, strictly speaking, the time between feeding of substrate and removal of digestate due to possible short-circuiting and can be determined by tracer analysis. MGRTs are commonly longer in plug-flow reactors.

To our current knowledge, anaerobic digestion process conditions are considered to be adverse to the exposed pathogens, particularly due to temperature effects and ammonia concentrations. Therefore, the longer the pathogens are exposed to the process conditions, the more intense is the resulting inactivating effect. Since short-circuiting shortens the given retention time and might thereby reduce the inactivation rate, it should be avoided for sanitation purposes in practice.

2.4 Inhibitory Process Compounds and Intermediates

Several compounds can negatively affect the viability of pathogenic organisms. It is known that high concentrations of intermediates and products of the anaerobic metabolism, e.g. fatty acids, alcohols and likewise sulfides and ammonia promote the sanitizing effect of anaerobic digestion. Ammonia is generated as a side product of the degradation of nitrogenous matter and is considered to be toxic to bacteria. Passive diffusion into the cells probably causes for instance proton imbalance and changes in the pH value and thereby impairs the energy metabolism. Enzyme inhibiting effects have also been proposed [28]. The release of ammonia is in turn influenced by temperature and pH value [20, 21, 28]. If sulfate is present, sulfide is built within anaerobic digestion by sulfate reducing bacteria. Sulfide is toxic to rotaviruses and several groups of bacteria [20, 28, 30].

Moreover, several light metal ions, heavy metals and organic compounds can accumulate in anaerobic digestion that exhibit inhibitory properties dependent on their concentration. Chen et al. [28] reviewed their inhibitory effect on the anaerobic digestion process, in general. However, most compounds may also be toxic to pathogens, because they affect cell wall and membrane structures or disrupt enzyme function. High concentrations of light metal ions lead to salt toxicity due to increased osmotic pressure. Heavy metals can bind to proteins or replace prosthetic groups. Organic compounds such as solvents or long chain fatty acids can accumulate in bacterial membranes and affect ion gradients or cause cell lysis [28]. For instance, Henry et al. [31] and Kunte et al. [32] reported on the toxicity of volatile fatty acids.

However, a diverse microflora can acclimate to a certain degree to a complex of influencing factors in the digester environment. Adaptation to inhibiting substances in anaerobic digestion was reported for instance for high ammonia, salt and oleate concentrations [33–35]. This allows for a well-functioning process microbiology at inhibitor concentrations that severely affect incoming pathogens and thereby supports sanitation.

2.5 *Input Concentration of Pathogens and Pre-treatment Steps*

The fraction of surviving pathogens is related to the concentration of pathogens in the feedstock. If high input concentrations occur, residual pathogens in the output digestate are more probable. Therefore, pre-treatment steps reducing germs in biogas plant feedstock contribute to the hygienic quality of digestate. Best practice ensiling already strongly reduces germs that are quite resistant to acidic environments such as EHEC and other intestinal pathogens that survive the passage through the extremely acidic human stomach [36, 37]. Bandte et al. [38] described significant reduction of the viability of fungal phytopathogens in ensiled compared to fresh sorghum crop material after anaerobic digestion.

A germ reducing pre-treatment is obligatory, if animal by-products (except for manure, digestive tract contents, milk and milk products, eggs and egg products and colostrum, if their sanitary status is not critical) are used as feedstocks for anaerobic digestion. According to Regulation (EC) No 1069/2009 [39] and Commission Regulation (EU) No 142/2011 [40], animal by-products are categorized into 3 groups based on their hygienic risk potential. Besides uncritical substrates, category 3 and 2 material is permitted to be treated in biogas plants after pasteurization and pressure sterilization, respectively. Utilization of category 1 material (e.g. zoo/circus animals, animals suspected of being infected by TSE, etc.) in biogas plants is not allowed.

Pathogens may survive longer in the fermenter, if they are embedded in coarse-grained substrates. Such particles represent a protecting matrix and may not be effectively digested within the process. Crushing substrates as a pre-treatment improves the digestion process in two ways: (I) by bettering nutrient accessibility and preventing technical problems such as blockage of pumps or pipes and (II) by homogenization of the digester content and breakup of coarse-grained substrates, which makes the pathogens accessible to the digestion process and renders possible their decay.

2.6 *Process Technology and Storage of Digestate*

The effectiveness of sanitation is determined by the procedural composition of the individual biogas plant. As described in Sect. 2.5 pre-treatment steps can contribute to sanitation by decreasing the input concentration of pathogen numbers. Various technical configurations of biogas plants are in use, each with their own specific contribution to the effectiveness of sanitation throughout the entire process.

Plug-flow reactors commonly feature prolonged MGRTs than stirred tank reactors due to the decreased risk of short-circuiting (see Sect. 2.3). A prolonged MGRT improves the germ reducing effect. Multi-stage digester configurations containing first- and second-step digesters connected in series are widely used. The reduction rate is thereby mathematically improved many times over compared to the digestion of the same volume in one compartment [41–43]. Post-treatment such as storage of digestate also contributes to a further reduction of pathogens. This was shown for the storage of raw manure [44–46], but also for digestate [38].

2.7 Pathogen Species

The resistance of individual pathogens determines their inactivation rate in the biogas process considerably. The decimal reduction time (often referred to as "T_{90}", or "D-value") is the time required to decrease to one-tenth of the initial concentration ($\log_{10}$ reduction). D-values are a useful tool to compare the parameters amongst each other concerning their ability to survive.

Some bacterial (e.g. *C. botulinum*, *C. perfringens* or some *Bacillus* spp.) and protozoan pathogens (e.g. *Cryptosporidium parvum*) are able to build resistant permanent forms like spores or cysts that can overcome adverse environmental conditions. Endospores are particularly less sensitive to e.g. heat, desiccation and

Table 1 Parameters positively influencing the efficiency of sanitation in anaerobic digestion

Parameter	Promoter of sanitation
Physical	• High process temperature • Long retention time • Reduced short-circuit currents • Comminution of coarse-grained substrates
Chemical	• Considerable changes in pH diverging from neutral • High concentration of intermediates and products of the anaerobic metabolism, e.g. ammonia, sulfide, fatty acids, alcohols, solvents, light metal ions or heavy metals • Low redox potential
Biological	• High microbial metabolic rate causing strong competition for food and predation [14] • Low initial concentration of pathogens (e.g. by germ reducing pre-treatment) • Acclimation of the biogas-producing microflora • Lack of adaptation of the pathogen to the digester environment - Low resistance to stressors - Disability to build permanent forms (e.g. spores)

chemical substances [47]. Spore- or cyst-forming organisms usually are characterized by prolonged D-values compared to the D-values of organisms that exist just as vegetative forms. Pasteurization, which affects mainly vegetative cells, is known to effectively reduce most non-spore-forming pathogens like *Salmonella* spp., *E. coli* pathovars or indicator organisms for sanitation. However, spore-forming and therefore less heat-sensitive bacteria (e.g. *Clostridium* spp. and *Bacillus* spp.) usually survive pasteurization almost uncompromised [48, 49]. They might even have a selective advantage thereafter if conditions meet their requirements.

The degree of inactivation depends on a multitude of factors. Table 1 shortly summarizes the impact factors discussed in this chapter.

3 Human and Veterinary Hygiene

3.1 Indicator Organisms for Biogas Processes

Depending on the different substrates used in the biogas process, different viral, bacterial and parasitic pathogens can enter the biogas production chain. The detection of some of these pathogenic agents is very extensive and complex. Therefore in routine diagnostic, the detection of bacterial hygienic indicator organisms frequently replaces the direct investigation of the presence of bacterial and also viral pathogens. As indicator organisms for the biogas process mainly *E. coli*, coliforms (e.g. [50]) and *Enterococcus* spp. are used.

Already in the early 1990s, Larsen et al. [51] tested the suitability of *E. coli* and *Enterococcus* spp. as indicator organisms or monitoring programs to control the hygienic status of digestate from large-scale biogas plants. It was shown, that *E. coli* as well as *Enterococcus* spp. were suitable as indicator organisms for monitoring the reduction of vegetative bacterial pathogens. However, the gram-positive enterococci are more stable toward environmental influences and the use of these bacteria as indicator organisms thus give the best safety margin [51]. These results have been confirmed by many other studies, e.g. with the study conducted by Watcharasukarn et al. [52]. Table 2 summarizes some published decimal reduction times for the two main indicator organisms for biogas processes, *E. coli* and *Enterococcus* spp.

The comparison of the data of different studies clearly shows that the decimal reduction time not only depends on the temperature conditions (mesophilic or thermophilic digestion) but also on the feedstock and other conditions of the biogas process, e.g. the ammonium concentration. Therefore, it is difficult to compare the decimal reduction times determined in different studies under special conditions or to apply them to other biogas processes. The determination of the content of indicator organisms thus plays an important role in the monitoring of the hygienic status of biogas digestate.

Table 2 Decimal reduction times (D-values) of *E. coli* and *Enterococcus* spp. in biogas processes using different feedstock (*n.d.* not defined)

Indicator organism	Feedstock	D-value			References
		At 37 °C	At 50–55 °C	At 70 °C	
E. coli	Fresh cow manure	1.92 d	9.15 min (55 °C)	0.03 min	Calculated according to [52]
	Biosolids	n.d.	10 min (55 °C)	n.d.	[53]
	Dairy manure	7-8 d	< 1 d (52.5 °C)	n.d.	[54]
Enterococcus spp.	Fresh cow manure	4.79 d	1.18 d (55 °C)	13.44 h	Calculated according to [52]
	Source-sorted organic fraction of municipal solid waste and egg albumin powder (high ammonia concentration)	0.8 d	n.d.	n.d.	[29]
	Source-sorted organic fraction of municipal solid waste	3.7 d	n.d.	n.d.	[29]

3.2 *Bacterial Pathogens in Biogas Processes Relevant for Human and Veterinary Health*

3.2.1 Bacterial Infectious Agents

Salmonella spp.

Salmonella spp. are the most common foodborne pathogens causing disease in humans and animals (salmonellosis). Animals are the main reservoir of *Salmonella* spp. although they rarely get infected. Farm animals such as cattle, pig and poultry represent the main source of contaminated food, but also of infested feedstock for anaerobic digestion (e.g. by-products, manure or dung). Feedstocks of animal origin can thus contaminate agricultural biogas plants with *Salmonella* spp. [55, 56].

A number of statutory regulations draw on *Salmonella* spp. in general, and on *Salmonella enterica* ssp. *enterica* sv. Senftenberg in particular, a heat-resistant representative of the genus *Salmonella*, as an indicator organism for sanitation and the evaluation of the hygienic status in diverse treatment procedures (e.g. Regulation (EG) No 1069/2009 [39] and Commission Regulation (EU) No 142/2011 [40]; German regulation for biological waste, BioAbfV [57], German regulation for realization of the law of animal byproduct removal, TierNebV [58], German regulation for fertilizers, DüMV [59]).

Salmonella spp. are inactivated by relatively short exposition at high temperature and were reduced completely at a pasteurization temperature of 70 °C [60, 61]. Several authors reported that decimal reduction of *Salmonella* spp. during anaerobic digestion occurs within the time range of a few minutes at thermophilic temperature and hours to a few days at mesophilic temperature. This is to some extent dependent on the heat resistance of the *Salmonella* sp. of concern. A Finnish study detected spiked *S.* Senftenberg in weekly analyses for 5 weeks during mesophilic (35 °C) anaerobic co-digestion of dairy manure and biowaste without pasteurization. With pasteurization at 35 °C or during thermophilic anaerobic digestion (55°), the detection occured only immediately after spiking and not thereafter [62, 63]. In laboratory-scale digesters digesting maize silage at 60 °C, *S.* Senftenberg were completely reduced by 8 orders of magnitude within 30 min, corresponding to a D-value of less than 3.2 min. Mesophilic digestion (38 °C) resulted in a reduction of at least 99.95 % after 48 h and a D-value of 8.3 h [8]. Olsen and Larsen [5] determined average D-values of 2.4 d and 2.1 d for anaerobic digestion of cattle or pig slurry at 35 °C and 0.7 h and 0.6 h at 53 °C for *S.* Typhimurium and *S.* Dublin, respectively. Smith et al. [60] found that decimal reduction of *S.* Senftenberg and *S.* Oranienburg within anaerobic digestion of centrifuged liquid raw sludge (biosolids) occurred in ca. 3 min and yet faster for S. Typhimurium at 55 °C. A 1.5–2 $\log_{10}$ reduction happened within 20 days at 35 °C [60]. Hoferer [9] determined D-values for *S.* Senftenberg in anaerobically digested cattle manure of 25.2 h at 35 °C and between 4.2–6.6 min at 55 °C.

***Campylobacter* spp.**

Campylobacter spp. are one of the most common bacterial causes of enteritis and belong to the group of zoonotic pathogens [6, 64, 65]. The organisms are ubiquitous in the environment and colonize a broad spectrum of animals as enteral commensals [64]. Pathogenic *Campylobacter* spp. are prevalent in biogas plant feedstock of animal origin [48, 66, 67]. Hutchison et al. [68] detected the organisms in 13, 14, 19 and 21 % of more than 1000 fresh samples of British cattle, pig, poultry and sheep manures, respectively. Lebuhn and Wilderer [12] determined low abundances in cattle manure ($\leq$10 genomes $*$ g^{-1} fresh mass) using a group-specific real-time PCR for thermophilic *Campylobacter* spp. (*C. jejuni, C. coli, C. lari* and *C. upsaliensis*). Higher abundances in the range of 10^3–10^4 cfu g^{-1} of *Campylobacter* spp. in raw pig or dairy manure were also reported [76, 77]. Pasteurization before anaerobic digestion effectively eliminated *C. jejuni* in substrates of animal origin [69]. Literature reports on the behavior of *Campylobacter* spp. concordantly ascertained its reduction in psychrophilic, mesophilic and thermophilic anaerobic digestion [10, 12, 70–77], though some studies found no change or nearly no reduction in *Campylobacter* concentration [78, 79]. Low abundances of thermophilic *Campylobacter* spp. were completely inactivated by anaerobic mono-digestion of cattle manure in a pilot-scale biogas plant at 55 °C [12]. Manyi-Loh et al. [77] recorded faster reduction of *Campylobacter* sp. by 90–99 % than of *E. coli* and a *Salmonella* spp.. In sentinel chamber experiments, Ade-Kappelmann [75] determined reduction of *C. jejuni* by ca. 2 orders of magnitude within 24 h at 35 °C and complete inactivation corresponding to a 5 $\log_{10}$ reduction after 1 h at 55 °C. Massé et al. [76] reported on 1.1–1.5 $\log_{10}$ reduction in anaerobic digestion of pig manure undergoing 7 d or 14 d of psychrophilic batch treatment. Knie et al. [10] determined a 4 $\log_{10}$ reduction of *C. jejuni* within 21 d in mesophilic (33 °C) and a complete inactivation (6–7 $\log_{10}$ reduction) within 24 h in thermophilic anaerobic digestion (55 °C) of manure and biowastes.

D-values describing the fate of *Campylobacter* spp. within the biogas process are comparatively rare and differ considerably from the D-values described for simple heat treatment (Table 3). In pure culture within culture bags exposed to an anaerobic digester at 35 °C, *C. jejuni* had a mean D-value of 3.6 d [80], but they were only slightly and very slowly reduced during anaerobic digestion at 28 °C [79]. Variation in data concerning the fate of *Campylobacter* spp. within anaerobic digestion is probably to be expected owing to the fact that this genus is able to convert into a viable but non-culturable (VBNC) state that makes cultivation by standard methods difficult [81–83].

Shiga toxin-producing/enterohemorrhagic *Escherichia coli* (STEC/EHEC)

Escherichia coli are non-spore-forming, gram-negative bacteria. Numerous strains are found as commensals in human or animal intestines, though some strains are pathotypes that can cause enteric/diarrheal illness [85]. These include: enteropathogenic *E. coli* (EPEC), enterohemorrhagic *E. coli* (EHEC), enterotoxigenic *E. coli* (ETEC), enteroaggregative *E. coli* (EAEC), enteroinvasive *E. coli* (EIEC) and diffusely adherent *E. coli* (DAEC). Among these, EHEC and EPEC (the latter causing infant diarrhea) are the most important pathotypes. EHEC, EPEC and ETEC can also cause disease in animals [85, 86].

Besides *Salmonella* spp. and *Campylobacter* spp., EHEC are one of the most common bacterial enteric pathogens in the western industrial countries and are also able to cause a life-threatening post infectious syndrome, the hemolytic uremic syndrome (HUS) [85, 87]. EHEC are capable of toxin production ("Shiga toxin", "Shiga-like toxin", "Verotoxin") and thus were also called Shiga toxin- or Verotoxin-producing *E. coli* (STEC/VTEC). Ruminants, particularly cattle, but also sheep and goats, were identified as main reservoir and carriers without symptoms [88–90].

Table 3 Decimal reduction times (D-values) for *C. jejuni* strains determined in lab and full-scale anaerobic digestion or by heat treatment

Process			D-value	References
Feedstock	Scale[a]	T (°C)		
Not specified	F	28	438.6 d	[79]
Not specified	L	35	3.6 d	[80]
Autoclaved biowaste sludge	B	50	237 min	[74]
Skim milk, beef, lamb or chicken	H	55-56	0.62–2.22 min	[84]

[a]*L* lab-scale, *F* full-scale, *B* batch experiment, *H* heat treatment

A German study screened 163 samples originating from 26 full-scale biogas plants in Bavaria. 26.4 % of the analyses resulted in the detection of EHEC. Positive results were obtained for 8.2, 52.9, 35.6 and 26.9 % of 49 plant substrates (e.g. silages), 17 animal substrates (e.g. manure), 45 contents of the main digester and 52 contents of second-step digesters, respectively [91]. The detection rates were high in this study and consistent with the fact that the excretions of ruminants are an important reservoir of EHEC [92]. Other recent studies revealed far lower detection rates for EHEC: Breves [93] found no EHEC bacteria in 105 substrate and digestate samples originating from 15 biogas plants in northern Germany fed with renewable resources and mixtures with manure. Bonetta et al. [94] analyzed the digestate of an Italian biogas plant fed with manure and agricultural by-products and did not detect EHEC O157:H7. A study in the USA measured low concentrations of EHEC O157: H7 in manure-fed laboratory-scale digesters [53]. Studies analyzing the behavior of EHEC in agricultural biogas processes specify differing data on decimal reduction times (Table 4).

Interestingly, D-values for *E. coli* serovars determined by Spinks et al. [95] in a water bath (55 °C) match almost perfectly to most of the D-values in thermophilic anaerobic digestion (Table 4) indicating that the unfavorable temperature is most relevant for the sanitizing effect in thermophilic biogas processes. However, other

Table 4 Decimal reduction times (D-values) for *E. coli* strains determined in lab and full-scale anaerobic digestion or water bath experiments

Strain	Process			D-value	References
	Feedstock	Scale[a]	T (°C)		
Indigenous	Dairy manure	B	25	9–10 d	[54]
E. coli O157	Cattle manure, food waste	F	35	27.6 h	[9]
O8; O147	Cattle or pig manure		35	1.8 d	[5]
Indigenous	Dairy manure	B	37	7–8 d	[54]
EPEC	Grass and maize silage	L	38	3.2–14.9 h	[91]
EPEC	Grass and maize silage	L	55	≤3.5 min	[91]
Wild-type	Dairy manure	B	55	≤10 min	[53]
O157:H7	Dairy manure	B	55	≤ 10 min	[53]
O157	Pig manure, food waste	L	55	1.2 min	[9]
O157	Cattle manure, food waste	L	55	1.8 min	[9]
O4; O147; O149	Cattle or pig manure	L	53	24 min	[5]
indigenous	Dairy manure	B	55	< 1 d	[54]
O3:H6	Pure culture	WB	55	6.7 min	[95]
O157:H7	Pure culture	WB	55	3.7 min	[95]
O157	Pure culture	WB	50	1.57 h	[9]
O157	Pure culture	WB	55	0.39 h	[9]

[a]*L* lab-scale, *F* full-scale, *B* batch experiment, *WB* water bath

experiments on *E. coli* O157 in water bathes resulted in far higher D-values [9]. The optimal growth temperature for *E. coli* O157:H7 was determined in tryptic soy broth at 37 °C and was similar for non-pathogenic *E. coli* serovars [96]. Reduction effects within mesophilic processes are therefore not expected to be caused by temperature alone but also by other factors of the digester environment (e.g. ammonia concentration, competition).

A previous study of Aitken et al. [53] at 55 °C resulted in similar D-values for a wild-type, non-pathogenic *E. coli* strain as well as for a slightly more heat-sensitive, putative O157:H7 pathovar. Aitken et al. [53] and Spinks et al. [95] showed that *E. coli* O157:H7 was comparably reduced and that the reduction behavior of wild-type *E. coli* strains is transferable to *E. coli* pathovars.

Yersinia spp. (*Y. enterocolitica* and *Y. pseudotuberculosis*)

The majority of *Yersinia* spp. is considered to be non-pathogenic to humans. Three species are relevant to human health and were also found in domestic and wild animals: (I) *Yersinia pestis*, the causative organism of bubonic and pneumonic plague that has been irrelevant in Europe for the last decades, (II) *Yersinia enterocolitica* that causes gastroenteritis and systemic infections and (III) *Yersinia pseudotuberculosis* that occasionally infects humans with a typhoid-like illness [97, 98]. *Y. enterocolitica* is one of the major pathogens present in animal manure [66, 99] and swine, the principal animal reservoir for pathogenic *Y. enterocolitica* [100, 101] The organism is thus considered as a risk factor if slaughterhouse waste or other animal by-products are used as biogas feedstock [6, 52]. A number of studies analyzed raw manure and digestates but did not detect *Y. enterocolitica* [12, 102–104] or *Yersinia* spp. [94].

Y. enterocolitica is sensitive to heat treatment. Drča [105] observed complete inactivation by pasteurization (70 °C, 1 h) of biogas substrate corresponding to 8 $\log_{10}$ reduction. At 50 °C, 1 $\log_{10}$ reduction was obtained within 20 min [105]. At 55 °C, D-values in milk were specified between 1.8 and 2.2 min [84]. At mesophilic conditions, studies on *Y. enterocolitica* resulted in D-values of 0.9 d in pure culture within culture bags and in 2.5–18.2 d in mixed cultures during semi-continuous anaerobic digestion [79, 80, 106]. Avery et al. [107] reviewed some studies and concluded that a reduction of *Y. enterocolitica* by 1.1–6 $\log_{10}$ units took place within 7–25 d during mesophilic anaerobic digestion. Massé et al. [76] detected *Y. enterocolitica* in 90 % of raw pig manure samples but proved a significant reduction during psychrophilic anaerobic digestion in batch systems.

Bacteria of the *Mycobacterium tuberculosis* complex and *Mycobacterium avium* ssp. *paratuberculosis*

Mycobacterium sp. belonging to the *M. tuberculosis* complex induce similar pathologies in different mammals [108]. Among them, *Mycobacterium tuberculosis* causes human tuberculosis and *Mycobacterium bovis* results in beef tuberculosis but infection of humans is also possible. Besides the *M. tuberculosis* complex species, *Mycobacterium avium* ssp. *paratuberculosis* (MAP) causes paratuberculosis (Johne's disease), a chronic enteric disease in domestic and wild ruminants and other animals [61, 109, 110]. MAP's causal involvement in Crohn's disease in humans is the subject of current discussion [111, 112]. MAP is abundantly released in the faeces of infested animals [109, 113, 114] and is therefore relevant concerning sanitary considerations in biogas plants. MAP is sensitive to heat treatment and was inactivated at 70 °C for 30 min [75]. Pre-treatment by pasteurization would thus sufficiently sanitize biogas feedstock. However, thermophilic as well as mesophilic anaerobic digestion also significantly reduces MAP [11, 113, 115, 116]. Olsen et al. [115] observed a 4 $\log_{10}$ reduction of MAP within 28 d and 3 h anaerobic digestion of bovine slurry in batch mode at 35 and 53 °C, respectively, and a complete reduction (>3 $\log_{10}$ units) in 24 h at 55 °C. Bendixen [11, 116] determined D-values of 0.7 h and 6 d at 53 °C and 35 °C, respectively. However, Slana et al. [114] analyzed MAP during anaerobic digestion in a full-scale biogas plant operated at 41–42 °C and detected viable organisms for 2 months after insertion of infested cow manure, whereas DNA was found for 16 month. Data on other *Mycobacterium* sp. than MAP are hardly available. In pyrosequencing approaches, 0.0025 % of the sequences retrieved from mesophilically digested biosolids in wastewater treatment plants were assigned to *M. avium* and none to *M. tuberculosis* [104]. 0.16 % of the sequences retrieved from a digestate after digestion of renewables and chicken manure at 41 °C were assigned to *Mycobacterium* sp. [117].

3.2.2 Spore-Forming Bacterial Pathogens

Among the spore-forming bacteria, particularly *Clostridium* spp. play an important part in the degradation processes that accompany biogas production. Besides their functional part, several *Clostridium* ssp. and *Bacillus* spp. can act as pathogenic agents causing disease basically by the formation of toxins. Among these, botulinum neurotoxin produced by *Clostridium botulinum*, is the most poisonous biological toxin known thus far [118].

Botulinum neurotoxin- (BoNT-)producing *Clostridium* spp.

The term BoNT-producing *Clostridium* spp. embraces the species *Clostridium botulinum* (groups I–III, toxin types A–F), *Clostridium argentinense* (=*C. botulinum* group IV, toxin type G) and BoNT-producing strains of *Clostridium butyricum* (toxin type E) and *Clostridium baratii* (toxin type F) [119]. Intoxication with BoNT causes botulism, a severe, life-threatening disease accompanied by paralysis [120]. *C. botulinum* and its spores are ubiquitous in soils (e.g. [121]) and can be found in faeces of livestock animals [122–129]. Other studies did not detect the organism in animal manure or dung and vegetable feedstock for biogas plants [130–132], but high input concentrations probably can occur if carcasses of small animals accidentally enter silages at crop harvest and ensiling [133].

Fröschle et al. [132] analyzed 43 sludges of main biogas digesters and 52 digestates and did not detect BoNT-producing *Clostridium* spp.. *C. botulinum* was neither found in lab-scale biogas reactors fed with cattle, pig and poultry manure and with renewable resources [134] nor in 105 substrates and digestates from 15 full-scale biogas plants in Germany [135]. Metagenomic analyses of digestates did also not indicate the presence of *C. botulinum* [136, 137]. However, Köhler [138, 139] detected *C. botulinum* occasionally in feedstock and digestates of biogas plants. The presence could be related to the use of moldy substrate. The fate of different pathogenic *Clostridium* spp. in a biogas plant digesting slaughterhouse waste was analyzed by Bagge et al. [131]. Thermophilic anaerobic digestion (52 ° C) led to 2.5 $\log_{10}$ reduction of pathogenic *Clostridium* spp.. The authors did not detect *C. botulinum* after anaerobic digestion of substrate containing the organism, indicating reduction of the pathogen. In sentinel chamber experiments, spiked *C. botulinum* cells were reduced with D-values between 1.0 ± 0.2 d at 55 °C and 34.6 ± 11.2 d at 38 °C [132].

In summary, the risk of encountering *C. botulinum* in digestate is very low if good agricultural practice is applied. Most importantly, there is no evidence yet for propagation of *C. botulinum* in biogas processes. On the contrary, the pathogen is reduced in biogas processes, although its reduction is obviously slow. Just like with other pathogens, the hygiene risk in digestate is thus lower than in feedstock.

Clostridium perfringens

C. perfringens is a ubiquitous spread spore-forming pathogenic bacterium and can be detected in soil or even in the human and animal intestine. It is a causative agent of a variety of human and animal diseases particularly dependent on the expressed toxin type (e.g. gas gangrene, foodborne illness, enterotoxaemia; necrotic enteritis) [49, 140].

Feedstock for biogas plants showed high incidence of *C. perfringens* i.e. in 9–95 % of the samples analyzed [48, 94, 135, 139, 141]. Findings concerning *C. perfringens* in digestates are divergent. In some community analyses, no evidence for *C. perfringens* was seen in digesters [136, 137], whereas other studies found the pathogen with in part very high incidence in anaerobic digestion i.e. 11–95 % of the samples analyzed [48, 94, 104, 135, 139, 141, 142]. Bagge et al. [48] analyzed four practice biogas plants treating pig or dairy manure and biowaste at mesophilic and thermophilic temperature and detected similar concentrations of *C. perfringens* before and after pasteurization, after anaerobic digestion and in storage tanks of 4.4 $\log_{10}$ cfu $*$ g^{-1} of fresh weight on average. Sahlström et al. [69] performed a heat treatment experiment in a water bath within biowaste and manure feedstock of Swedish biogas plants. *C. perfringens* was not significantly affected by 55 °C or 70 °C for 30 or 60 min leading to the conclusion that pasteurization (that is dictated along with anaerobic digestion of animal by-products of category 3 by Regulation (EC) No 1069/2009 [39] is not sufficient for the elimination of *C. perfringens*.

Published results on the fate of *C. perfringens* in anaerobic digestion are divergent. A number of studies ascertained no reduction of this pathogen at phsychrophilic, mesophilic or thermophilic conditions [5, 11, 72, 76, 143]. Contrary to that, some studies discovered reduction of *C. perfringens* in mesophilic and thermophilic anaerobic digestion [10, 52, 74, 144]. Watcharasukarn et al. [52] reported reduction of *C. perfringens* in anaerobic digestion of cow manure by 1.35 $\log_{10}$ units at 37 °C (15 days; D-value: 4.05 d) and <1 $\log_{10}$ reduction at 55 and 70 ° C (after ca. 3 h and ca. 30 min, respectively, no D-value calculable). The reduction behavior appears to be predominantly determined by the ability of *C. perfringens* to build spores and the divergence of results published might be explained by different percentages of spores and vegetative cells applied to the analyses.

Other *Clostridium*spp. relevant for human and veterinary health

C. novyi type A, B and D (the latter is also referred to as *C. haemolyticum)* can cause infectious diseases (e.g. infectious necrotic hepatitis, edema, wound infections in animals and gas gangrene in humans), whereas *C. novyi* type C is non-toxic and apathogenic. *C. chauvoei* causes blackleg, a gas gangrenous infection in cattle and sheep and *C. septicum* is the reason for braxy and malignant edema and gas gangrenous wound infections [140, 145]. *C. difficile* is an emerging nosocomial pathogen of increasing relevance worldwide causing antibiotic associated diarrhea and colitis [146].

Reports on these *Clostridium* spp. in agricultural feedstock and biogas plants are scarce. A screening regarding German biogas feedstock and digester contents resulted in detection of *C. novyi* in 2.3 % of 44 plant feedstock samples, none of 17 animal feedstock samples, 7.1 % of 42 contents of main digesters and 3.9 % of 51 digestates. The concentration in digester contents was less than 3 cfu $*$ g^{-1} of fresh weight. *C. haemolyticum, C. septicum* and *C. chauvoei* were not found in 154 samples [132]. Bagge [49] did not find *C. haemolyticum* and *C. chauvoei* in mesophilic and thermophilic Swedish biogas plants fed with manure and biowaste after pasteurization. *C. septicum* was found in the substrates of 2 of 4 analyzed biogas plants before and after pasteurization. *C. haemolyticum* was completely reduced within a few days in spiking experiments [49]. *C. chauvoei* was detected in 3 of 11 feedstock samples for biogas plants but not after pasteurization and after the following anaerobic digestion. The organism was not found in soil samples (n = 84) and silage samples (n = 4) and only in 1 manure sample (n = 114) [147]. *C. difficile* was detected in 25 % or 38 % of analyzed biosolids after mesophilic and mesophilic/thermophilic temperature phased anaerobic sewage treatment [148]. Metagenomic approaches repeatedly detected contigs similar to *C. difficile* in digesters, but reliable identification could not be achieved [137, 149].

Knowledge on *Clostridium* spp. in agricultural biogas production is scarce. Further research is required particularly on *C. difficile,* showing high incidence in anaerobic digested sewage sludge.

Bacteria of the *Bacillus cereus* group

The *Bacillus cereus* group contains several *Bacillus* spp. that are phylogenetically highly related. This group includes *Bacillus anthracis*, *Bacillus cereus*, *Bacillus* *thuringiensis*, *Bacillus mycoides*, *Bacillus pseudomycoides* and *Bacillus weihenstephanensis*. *B. thuringiensis* is an insect pathogen. *B. cereus* is associated with regressive relevance with human food poisoning causing diarrhea and abdominal distress or nausea and vomiting and occasionally with mastitis in cattle. *B. anthracis* is the causal agent of anthrax affecting particularly herbivores but also mammals including humans. *Bacillus* spp. can be found e.g. in soil [150, 151] and in biogas feedstock. Vissers et al. [152] detected a mean concentration of 2.4 ± 0.07 $\log_{10}$ *B. cereus* spores $* \, g^{-1}$ silage. *B. thuringiensis* and *B. cereus* were detected in slaughterhouse waste, *B. cereus* additionally in cattle manure used as biogas feedstock [131]. 1.4 % of all cultivable isolates from substrates belonged to *B. cereus* group [141].

As *Bacillus* spp. are able to form endospores, they are comparatively resistant to environmental impacts such as heat (D-values of up to 36 min at 95 °C) [84], pressure and chemical agents and thus also to the sanitizing effect of biogas processes. Several studies found no or almost no reduction of *Bacillus* spp. [131], *B. cereus* group [12, 72, 153] or *B. cereus* [5, 11] in mesophilic and thermophilic biogas processes. Thus several proofs of *B. cereus* or of *B. cereus* group exist for digestates in abundances of $10^2 - 5 \, * \, 10^3$ cfu $* \, mL^{-1}$ [12, 117, 131, 141, 154]. *Bacillus* spp. even pass unaffectedly through pasteurization [131].

3.3 Pathogenic Parasites in Biogas Processes

Just like bacterial pathogens, potentially pathogenic human and animal parasites can be introduced into the biogas process especially through contaminated substrates of animal origin like slurry or manure. Since, parasites usually need a host organism for growth and proliferation there is no indication for multiplication of these organisms in the biogas process. Therefore, the sanitary status of digestate is assessed by the viability of cysts/oocysts (protozoa) or eggs (e.g. nematoda) in the biogas processes or treatment steps.

3.3.1 Protozoa

The most important protozoa potentially transmitted by biogas feedstock and digestate are *Cryptosporidium* spp. and *Giardia* spp.. Both genera contain pathogens causing gastrointestinal illness in humans and are also infective to a variety of animals. Cattle and domestic animals are considered as main reservoir [97, 155–157]. A study from the United Kingdom showed the occurrence of both parasites in cattle manure in maximum concentrations of 10^3 organisms $* \, g^{-1}$ [158]. Thus, an input into the biogas process via contaminated manure or slurry can be assumed.

Côté et al. [103] observed a decrease of indigenous *Cryptosporidium* spp. and *Giardia* spp. in swine manure to undetectable levels within 20 d in psychrophilic anaerobic digestion. Sentinel chamber experiments in anaerobic digestion of cattle manure resulted in a reduction of viable counts of *C. parvum* oocysts of >1 $\log_{10}$ unit during 4 h at 38 °C and of ≥2 and ≥5 $\log_{10}$ units during 4 h at 55 °C whereas almost no reduction was seen in real-time PCR assays on the DNA level [12, 159]. This goes in line with the results of Godfree and Farrell [71]. Chauret et al. [160] obtained no statistically significant reduction of both parasitic permanent forms during mesophilic anaerobic digestion of sewage sludge, but could not distinguish between viable and non-viable forms.

However, studies on cryptosporidia and giardia in other matrices than manure and slurry indicate that their permanent states (oocysts and cysts, respectively) are only fairly tolerant to heat [161]. Inactivation in water amounting several orders of magnitude is thus possible after 10 min at 70 °C [162, 163]. Therefore, biogas feedstock of animal origin should be sanitized using heat treatment at temperatures not less than 70 °C in order to achieve a reliable and complete inactivation of protozoa cysts/oocysts. Alternatively, thermophilic treatment with a longer retention time is an option. Without heat treatment, a residual fraction of cysts might survive the biogas process.

3.3.2 Helminths

Besides cysts/oocysts of protozoa, eggs of helminths can pose a risk for human or animal health. Although helminth eggs are not as heat-stable as the mentioned protozoa cysts, some species could survive extreme pH values [161]. Therefore the eggs of *Ascaris suum* were used as indicator organism for testing the survival of helminth eggs in the biogas process in some studies [69, 164]. In the study of Plym-Forshell [164] 60 % of the *A. suum* eggs survived at temperatures of 22–27 °C up to 56 days. In both studies however, *A. suum* eggs were inactivated at temperatures of 55 °C. Accordingly, thermophilic anaerobic digestion appears to be sufficient to inactivate helminth eggs. However, mesophilic anaerobic digestion cannot guarantee a complete reduction of viable helminth eggs. To ensure a hygienically safe digestate, substrates of animal origin should be heat-treated before application to mesophilic anaerobic digestion.

3.4 Viral Pathogens in Biogas Processes

Unlike bacteria, viruses are not able to multiply outside of a host organism. Two factors are thus of major interest for the sanitary status of the fermentation substrates: the amount of viruses in the substrates used in the biogas process and the reduction capacity of different process technologies concerning the different viruses.

The ability to survive depends decisively on the different virus structures. In general, enveloped virus families such as *Herpesviridae, Poxviridae* and *Flaviviridae* are much more sensitive to environmental conditions than uncoated virus families (e.g. *Parvo-*, *Picorna-*, *Calici-* and *Circoviridae*). Nevertheless, resistance to different environmental conditions like heat stability can also strongly differ between the different virus families in the group of uncoated viruses. Representatives of *Parvo-*, *Calici-* and *Circoviridae*, for instance, are more heat-stable than those of *Poxviridae* [165]. The most important factors inactivating viruses are heat treatment, high and low pH and radiation. Furthermore decomposition can cause inactivation especially of enveloped viruses [165]. Table 5 shows a summary of guideline values for the stability of enveloped and uncoated viruses with regard to environmental conditions important in the biogas process.

The guideline values in Table 5 show that it is nearly impossible to predict an inactivation rate for different virus families in different biogas processes. The inactivation rate not only depends on temperature and pH but also on the composition of the biomass, which varies not only between the different types of biogas plants but also in the same biogas plant depending on the substrate and the composition of the substrates. Therefore, it is not only very difficult to compare data from different studies with regard to the inactivation of viruses in biogas plants, but also to transfer the results from in vitro experiments to conditions experienced in vivo. However, the guideline values in Table 5 show that only uncoated viruses should be used as indicator organisms, because inactivation conditions from experimental studies using uncoated viruses can also be applied for enveloped viruses. The inactivation capacity of different biogas processes is therefore typically assessed using two different uncoated viruses as indicator organisms, mainly a representative of the family *Parvoviridae*, which is characterized by a high thermal resistance and a more heat-sensitive virus like a representative of the family *Picornaviridae* [166].

Table 5 Summary of guideline values for the stability of enveloped and uncoated viruses (according to [165])

Environmental conditions		Inactivation/stability of	
		Enveloped viruses	Uncoated viruses
Heat treatment	22° C	Hours to days (according to the virus and other environmental conditions)	Days to weeks (according to the virus and other environmental conditions)
	37 °C	Hours to days (according to the virus and other environmental conditions)	
	56 °C	Minutes to hours (according to virus and other environmental conditions)	
	80 °C	Minutes (with exception of some uncoated virus families e.g. *Parvoviridae*, which persist for hours)	
	100 ° C	Seconds	
Low pH	pH 2	Predominantly unstable	Partially unstable
High pH	pH 13	Unstable	Relatively unstable
Decomposition		To varying degrees unstable	Relatively stable

3.4.1 Enterovirus

The genus *Enterovirus* belongs to the family of *Picornaviridae* and includes important veterinary viral pathogens like *Enterovirus* E (formerly bovine enterovirus) and Swine vesicular disease virus (SVDV). These viruses can enter the biogas process with contaminated substrates, e.g. manure or slurry of subclinically infected animals [69].

In in vitro studies, representatives of the genus *Enterovirus* were reduced in digestates by max. 1.6 $\log_{10}$ units using a temperature of 55 °C for 30 min and 6.4 $\log_{10}$ units using a temperature of 70 °C for 30 min [69]. Lund et al. [166] showed that a minimum guaranteed retention time (corresponding to a 4 $\log_{10}$ reduction) of 23 h at 35 °C (mesophilic anaerobic digestion) and <0.5 h at 55 °C (thermophilic anaerobic digestion) is necessary for *Enterovirus* in in vivo studies using laboratory-scale biogas reactors. These results correspond with results of other studies, e.g. [167]. In this study a human Coxsackievirus was used as a model for the genus *Enterovirus*. After four hours of thermophilic digestion no infectious virus was detectable and after 2 days also no virus RNA was found in the samples (7 $\log_{10}$ reduction).

3.4.2 Parvovirus

The family *Parvoviridae* comprises more heat-resistant representatives of the group of uncoated viruses, like the bovine parvovirus (BPV) and the porcine parvovirus (PPV). Just like enteroviruses, these viruses can enter the biogas process chain via contaminated slurry or manure [69]. Prado et al. [168] found that norovirus, a representative of the family *Caliciviridae*, was present in 50 % of the investigated activated sludge samples. Activated sludge is commonly post-treated at wastewater treatment plants using anaerobic digestion. Furthermore, representatives of the family *Parvoviridae* can be used as indicator organisms for other virus families with a comparable heat resistance like *Calici-* and *Circoviridae*.

It has been shown in in vitro studies that representatives of the family *Parvoviridae* can survive the sanitizing conditions of pre-treatments used before anaerobic digestion. For example, Sahlström et al. [69] detected a mean reduction of only 2.6 $\log_{10}$ units after 60 min heat treatment at 70 °C, compared to a 6.4 $\log_{10}$ reduction for the genus *Enterovirus* using the same treatment. In in vivo studies, Lund et al. [166] determined a biphasic inactivation with a high initial reduction followed by a decreasing reduction rate. Therefore a minimum guaranteed retention time (corresponding to a 4 $\log_{10}$ reduction) of 11–12 h was necessary at 55 °C in the initial phase and 54 h thereafter to inactivate different representatives of the family of *Parvoviridae*. It can therefore be assumed that the more heat-resistant virus families like *Parvoviridae*, *Calici-* and *Circoviridae* can survive the conventional mesophilic and maybe also a thermophilic biogas process and can be found in digestate.

4 Phytosanitary Aspects

Regarding global concerns about the impacts of feeding infected biomass to biogas plants, phytopathogenic microorganisms are of particular significance as they are responsible for crop losses and interfere with food security. The current interest in phytosanitation was also stimulated by the ability of seeds or diaspores such as from various unwanted herb (weed) species to survive the passage through a biogas plant. In addition, promotion of feedstock diversity has received attention in the specialist community. The introduction of novel or non-native dedicated energy crops is associated with the potential for these crops to escape and invade outside of cultivation. Hence, there is a growing demand for scientifically sound and process-based studies to provide clear information about the efficiency of phytosanitation within the biogas production chain as well as the major factors and mechanisms contributing to the phytosanitary status of digestate.

The biogas production chain commonly comprises the steps supply (cultivation, harvest, ensiling), storage, pre-treatment, feeding and anaerobic (co-)digestion as well as digestate storage until spreading as fertilizer. In general, the incoming material represents the main source of pathogens and pollutants in digestate. The use of contaminated digestate as fertilizer can pose a phytosanitary risk if phytopathogens, seeds or vegetative propagules survive the biogas production chain.

4.1 Phytosanitary Effects of the Biogas Production Chain

According to Plöchl et al. [41], knowledge of fate and behavior of plant pathogens in the biogas production chain is limited and hampers the estimation and evaluation of potential phytosanitary risk. Reasons for this are manifold, including:

- the broad spectrum of phytopathogenic microorganisms (bacteria, viruses, fungi) and the occurrence of both infectious propagules and resting structures
- limits of sampling as well as of routine microbiological analytical techniques regarding (i) the heterogeneity of matrices and the microbial composition, (ii) identification and quantification of pathogens and (iii) reliability of lab-scale tests (in vitro and on rich culture media) for in vivo assessments
- higher investments for recent, more sensitive, accurate, specific and much faster diagnostic techniques (molecular-based techniques, real-time PCR) than conventional approaches.

In order to prevent the exposure of hazardous organisms to the environment, lab-scale experiments should be a prerequisite prior to perform investigations in commercial biogas plants. Within the framework of the comprehensive joint project "Investigations regarding the phytosanitary risk of anaerobic digestion of crop products in biogas plants" [169], systematic experiments were conducted regarding the effect of the mesophilic anaerobic digestion in continuously stirred tank reactors

(CSTR) on various host-pathogen-combinations. Fresh and ensiled plant materials, infected with specific fungal plant pathogens, were examined in germ carriers under variation of the exposure time in the digester (6–138 h) and the storage time of the digestate (4 weeks; 6 months) regarding inactivation and survival capability. Results of selected host-pathogen-combinations obtained in this way provide information whether and to which extent sanitation took place. These findings are presented in the following sections.

4.1.1 Plant Pathogen Species

Rodemann et al. [170] and Bandte et al. [38] reported that mesophilic anaerobic digestion of infected plant material led to considerable reduction or even complete inactivation of the plant pathogens tested. Required exposure times tended to be significantly longer in biogas plants than in small CSTRs (Table 6).

Sanitation of crop material infected with either *Sclerotinia sclerotiorum* or *Rhizoctonia solani* occurred within an exposure time of 6 h. These findings match those of Seigner et al. [171] who ascertained an exposure time of 8 h to inactivate the same two pathogens in lab-scale reactors operated at 38 °C. In contrast, sclerotia of *Sclerotinia cepivorum* were at least in part viable when recovered from an experimental reactor after 6 weeks of anaerobic composting [172].

Sanitation of other pathogens such as species of *Fusarium* was reached within 138 h (Table 6). These differences probably result from (i) the colonization type by the fungal pathogens and (ii) the consistence of the infected feedstock. *Fusarium proliferatum* and *Fusarium verticillioides* colonize the plant endogenously, while *S. sclerotiorum* lives on the surface of the plant. The infected sorghum plant material that was inserted into the digester was derived from field plots and contained more lignin due to its advanced physiological age. *Fusarium* spp. propagules were thus longer protected from degradation and inactivation than *S. sclerotiorum*. Prolonged exposure time resulted in strong reduction of infectious *Fusarium* spp. propagules (Table 6).

Bandte et al. [38] demonstrated that different species of a genus can be affected differently by anaerobic digestion. Unlike *F. proliferatum*, most of *F. verticillioides* propagules did not survive anaerobic digestion for more than 24 h. *Fusarium graminearum* decay was attained within the first 24 h in lab-scale reactors operated at 38 °C [171]. In those tests, pure *F. graminearum* cultures grown on wheat grains were exposed, in contrast to the investigations from Bandte et al. [38] in which infested whole plant biomass was used. The viability of fungal propagules in pure cultures tends to be shorter because of the absence of surrounding plant tissue that protects them from physical, enzymatic or chemical degradation. Rodemann et al. [170], for example, exposed species of *Fusarium*-infested maize plants to anaerobic digestion. They showed that *Fusarium culmorum* und *F. verticillioides* were inactivated within 24 h in lab-scale reactors.

It is concluded that the sanitation potential of the anaerobic digestion process is mainly determined by the pathogen species and not by the crop species.

Table 6 Proportion of aliquots with viable phytopathogens in different host crops after particular time periods exposed to mesophilic anaerobic digestion in either lab-scale reactors (at 37 ± 1 °C) or a full-scale biogas plant (at 40 ± 1 °C)

Pathogen	Host[a]			Scale[b]	Abundance (%) after time periods (h)[c]							References[d]
					0	6	24	48	72	96	138	
Claviceps purpurea	Rye	Ergot	f	F	54	27	0	0	0	0	0	[170]
Fusarium avenaceum	Cereal	Whole plant	f	L	60	10	0	n/d	n/d	n/d	0	[170]
	Maize	Whole plant	f	L	81	4	0	n/d	n/d	n/d	0	[170]
Fusarium culmorum	Maize	Grain	f	F	100	100	n/d	n/d	n/d	n/d	0	[170]
		Grain broken	f	F	100	100	n/d	n/d	n/d	n/d	0	[170]
		Whole plant	f	L	100	100	0	n/d	n/d	n/d	0	[170]
			f	F	80	30	3	0	0	0	0	[170]
	Wheat	Grain broken	f	L	100	7	0	n/d	n/d	n/d	n/d	[170]
		Grain	f	L	100	36	0	n/d	n/d	n/d	n/d	[170]
		Grain broken	f	F	100	99	0	0	0	0	0	[170]
		Grain	f	F	100	76	3	0	0	0	0	[170]
Fusarium proliferatum	Sorghum	Whole plant	f	L	100	100	87	0	0	0	0	[38]
			f	F	100	n/d	30	0	13	5	0	[38]
			e	L	100	0	3	0	0	0	0	[38]
			e	F	100	n/d	12	2	n/d	n/d	n/d	[38]
Fusarium verticillioides	Cereal	Whole plant	f	L	62	15	0	n/d	n/d	n/d	0	[170]
	Sorghum	Whole plant	f	L	100	53	1	0	0	0	0	[38]
			f	F	100	98	4	32	0	n/d	0	[38]
			e	L	60	0	0	0	0	0	0	[38]
			e	F	100	75	n/d	15	0	0	0	[38]
	Maize	Grain	f	F	70	70	n/d	n/d	n/d	n/d	0	[170]
		Whole plant	f	L	86	30	0	n/d	n/d	n/d	0	[170]
			f	F	62	51	4	0	0	0	0	[170]

(continued)

Table 6 (continued)

Pathogen	Host[a]			Scale[b]	Abundance (%) after time periods (h)[c]							References[d]
					0	6	24	48	72	96	138	
Rhizoctonia solani	Maize	Whole plant	f	L	73	12	0	n/d	n/d	n/d	0	[170]
			f	F	85	2	2	0	0	0	0	[170]
	Potato	Tuber	f	L	80	0	0	0	0	0	0	[38]
Sclerotinia sclerotiorum	Sugar beet	Tuber	f	L	100	0	0	0	0	0	0	[38]
	Sugar beet	Tuber	f	F	100	0	n/d	n/d	n/d	n/d	n/d	[38]

[a]Host crop material: *f* fresh, *e* ensiled
[b]*L* lab-scale reactor, *F* full-scale biogas plant
[c]*n/d* not determined

4.1.2 Ensiling of Biogas Crops

Pre-treatment of crop material influences the degree of pathogen depletion in the biogas production chain as well. Chopping at harvest and compaction during ensiling result in a lower polymerization of cell wall constituents which can be regarded as a kind of mechanical treatment [173]. Ensiling itself involves production of organic acids and a decrease in pH down to 4 that consequently prevents growth of fungi, yeasts and bacteria which may otherwise decompose the biomass. Bandte et al. [38] used ensiling as a biological pre-treatment method. In this study, the storage period after ensiling directly corresponded to the inactivation of *Fusarium* spp. during anaerobic digestion. Ensiling of crop material led to increased inactivation of *F. proliferatum* and *F. verticillioides* (Table 6). Significant reduction of *F. proliferatum* viability was observed in silage stored for 35 days compared to fresh sorghum during anaerobic digestion for 24 h [38]. *F. proliferatum* remained thus infectious in fresh sorghum plant material in about one-fifth of aliquots. The incorporation of ensiled sorghum decreased this portion to less than one-tenth of the original load. This is important as almost 90 % of plant material entering biogas plants is ensiled [2].

4.1.3 Anaerobic Digestion of Infected Plant Material

In order to understand the fate and behavior of plant pathogens in the biogas production chain, Plöchl et al. [41] systematized the available information focusing on the sanitizing impact of anaerobic digestion on particular phytopathogen host combinations. Therefore, simulation was used to demonstrate the effects influencing the steady state of pathogen infected plant material in both digesters and digestate. Simple kinetic inactivation approaches and mass balances of infected material were carried out considering single-step as well as two-step digestion.

D-values for inactivation gained from the experiments summarized in Table 6 ranged between almost null (i.e. spontaneous complete inactivation) and 96 h. Simulations of mass balances of infected feedstock and concentration in contaminated output were performed for pathogen-sorghum combinations. These combinations cover almost the entire range of D-values determined. The simulation revealed a “very fast” to “fast” reduction of infected material after a singular feeding, reaching a cutback to less than 1 % of input within 4 days even for D-values of 68 h. Steady state mass balances below input rate could be calculated with D-values of less than 2 h at a continuous hourly feeding. At higher D-values steady state mass balances exceeded the input rate but were still clearly below the sum of input mass. Dilution further decreased mass balances to values 10^{-5} to 10^{-6} $Mg * m^{-3}$ for first-step digestion and 10^{-8} to 10^{-9} for second-step. Hence, Plöchl et al. [41] could demonstrate the combined effect of dilution and reduction on the concentration of pathogenic propagules both in the digesters and the digestate.

4.1.4 Storage Period of Digestate

According to Bandte et al. [38], storage of digestate accounts for a further significant reduction in viable *Fusarium* spores in case of processing fresh sorghum feedstock. A storage period of 4 weeks already resulted in complete inactivation of *F. proliferatum* in sorghum being exposed for 6 h. After storage of 6 months, none of the digestates harbored viable *Fusarium* spp. propagules. Comparable results were gained with *Fusarium* spp. in infected ensiled sorghum. The exposure time required for a complete inactivation could be reduced to 6 h when digestates were stored for 4 weeks. No additional effect of storage was visible in regard to inactivation of *S. sclerotiorum* in sugar beet and *R. solani* in potato as the pathogens lost their infectivity in all sample carriers already after the minimal exposure time of 6 h.

4.2 Weeds and Alien Plant Invaders

As with plant pathogens, literature surveys reveal that studies on these subjects are scarce and fragmentary. According to Westerman et al. [174], using digestate as fertilizer, the probability of dispersal will be highest for (i) common weed seeds, as these have the highest likelihood of entering biogas plants, and (ii) seeds that are resistant to the adverse conditions in anaerobic digesters. Following their argumentation the risk will be highest for invasive, quarantine and troublesome plant species that do not have a widespread distribution yet.

Focusing on weed seeds, Westerman et al. [175] conducted experiments in lab-scale mesophilic batch reactors, either with or without ensiling. The experiments demonstrated that species with hard seeds were more likely to survive ensiling (up to 98 %) and mesophilic anaerobic digestion (up to 58 %) compared with species whose seeds lack a water impermeable layer (≤1 %). Findings reveal considerable variation in survival probability between and within species, and between silages and reactors. For hard-seeded species, survival depended on the proportion of hard seeds and on the ability to remain hard-seeded during treatment. Freshly harvested seeds lost their water impermeability and survived poorly. Regarding tomato seeds, which are used as indicator species in the sanitation of composts, the survival was comparable with that of the best surviving species without physical dormancy, but was not comparable with that of hard-seeded species. Hence, tomato can only represent species that are not hard-seeded.

In a corresponding experiment, Westerman et al. [174] tested the ability of seeds from five plant species to survive commercial biogas plants compared with results obtained from lab-scale reactors [175]. Seeds were exposed for 1–9 days in two large-scale commercial mesophilic biogas plants. The decimal reduction time (D-value) was estimated at 1.5 days for *Abutilon theophrasti* population D2003, 2 days for *A. theophrasti* population ES2008, 5.8 days for *Malva neglecta*, 4.7 and 19.7 days for *Chenopodium album*, and 1.2–9.1 days for *Fallopia convolvulus*. Regarding the differences in ranking of plant species, authors concluded that lab reactors are not necessarily a good model system for commercial reactors. Seeds could survive passage through biogas plants, although at (extremely) low numbers. If the objective is to eliminate all risks and dispose of all weed seeds, it will be necessary to either add extra sanitation steps, or to identify and manipulate the factors that are responsible for high seed mortality during anaerobic digestion. As a result of a recent pioneering review, Westerman and Gerowitt [176] concluded that it is currently unknown what these factors are.

In order to prevent further introduction and to reduce existing populations of alien (invasive) species, research on control methods is going on in Europe (e.g. EU project HALT Ambrosia [*Ambrosia artemisiifolia*]; [177]). In this context, the biogas process is considered as an alternative to incinerating the plant material escaping mechanical control measures such as cutting or uprooting. Whether this approach is a safe and environmentally friendly method of waste disposal or not must always be clarified case-by-case. Mesophilic and thermophilic anaerobic digestion of seed-carrying *A. artemisiifolia* plants is not a safe option [178]. In the case of common ragwort (*Senecio jacobaea*) which is developing into a serious pest weed in Northern Germany scientifically sound and process-based studies are still missing.

5 Conclusion

The sanitizing potential of anaerobic digestion in the context of agricultural biogas production depends on a multitude of factors. Transferability of results on specific or indicator organisms and similarly the application of lab-scale results to full-scale processes is limited. A generalizing estimation of the sanitizing effect is almost impossible as the extent of sanitation is particularly related to

(1) the organism of concern, especially its structure and ability to form permanent states (e.g. formation of spores or cysts/oocysts etc., kind of propagules, hard-seeded and not-hard-seeded weed species, enveloped or uncoated viruses, gram-positive or gram-negative bacteria) and
(2) the available process technology, the feedstock composition and last but not least the diligence of the operation management (storage and consistence of manure, quality of silage, kind and consistency of substrates, pre-treatment steps, homogenization, temperature conditions, stirring/short-circuit currents, one-stage/multi-stage processes, stability of the biogas biocenosis, storage of digestate, cleanliness, recontamination/black-white-separation, etc.).

Sanitation by anaerobic digestion is a beneficial side effect of the biogas process, but it is not its intrinsic function. The technology has not been developed to completely eliminate pathogens, and therefore, a hygienically safe status of digestate cannot be guaranteed. If the objective is to eliminate all risks and inactivate all contaminations, additional sanitation steps such as steam/pressure sterilization are necessary and in some cases mandatory.

However, the consensus of studies published on this issue is that the sanitary quality of digestate is improved compared to the untreated feedstock concerning non-spore-forming bacteria, pathogenic parasites, viruses, phytopathogenic fungi and weed seeds. In the worst case, the sanitary quality remains nearly unchanged (e.g. spore-forming bacteria of the genus *Bacillus* and certain clostridia). The spreading of digestate from agricultural biogas production in accordance to current rules and best practice recommendations is thus considered to elicit no additional danger for the health of humans, animals and plants.

References

1. Lebuhn M, Munk B, Effenberger M (2014) Agricultural biogas production in Germany—from practice to microbiology basics. Energy Sustain Soc 4:10. doi:10.1186/2192-0567-4-10
2. Möller K, Müller T (2012) Effects of anaerobic digestion on digestate nutrient availability and crop growth: a review: digestate nutrient availability. Eng Life Sci 12:242–257. doi:10.1002/elsc.201100085
3. Ziemba C, Peccia J (2011) Net energy production associated with pathogen inactivation during mesophilic and thermophilic anaerobic digestion of sewage sludge. Water Res 45:4758–4768. doi:10.1016/j.watres.2011.06.014
4. Morar MV, Oechsner H, Dancea Z, Kusch S (2009) The anaerobic digestion as process step for hygienisation of liquid manure and agro-industrial waste. In: Proceedings of the 8th international symposium "Prospects for the 3rd Millenium Agriculture." Cluj-Napoca, Romania, 7–10 Oct 2009, pp 149–154
5. Olsen JE, Larsen HE (1987) Bacterial decimation times in anaerobic digestions of animal slurries. Biol Wastes 21:153–168. doi:10.1016/0269-7483(87)90121-2
6. Sahlström L (2003) A review of survival of pathogenic bacteria in organic waste used in biogas plants. Bioresour Technol 87:161–166
7. Umetsu K, Kikuchi S, Nishida T, et al (2009) The effect of anaerobic digestion in biogas plants on survival of pathogenic bacteria. In: The organizing commitee of OASERD-APEID (eds) Education for sustainable development (ESD) on agriculture and livestock production and global environmental issues, pp 30–40
8. Fröschle B, Lebuhn M (2012) Abtötung von Salmonellen im Biogasprozess - LfL-Information. Bayerische Landesanstalt für Landwirtschaft (ed), Freising, Germany, http://www.lfl.bayern.de/mam/cms07/publikationen/daten/informationen/p_44803.pdf. Accessed 10 Feb 2015
9. Hoferer M (2001) Seuchenhygienische Untersuchungen zur Inaktivierung ausgewählter Bakterien und Viren bei der mesophilen und thermophilen anaeroben alkalischen Faulung von Bio- und Küchenabfällen sowie anderen Rest- und Abfallstoffen tierischer Herkunft. Inaugural dissertation, Freie Universität Berlin, Berlin, Germany
10. Knie A, Haumacher R, Philipp W, et al (2001) Untersuchungen zur Seuchen-und Phytohygiene in Anaerobanlagen (Halb-bzw. Großtechnische Anlagen). Forschungsbericht FZKA-BWPLUS, Universität Hohenheim, Hohenheim, Germany. http://www.fachdokumente.lubw.baden-wuerttemberg.de/servlet/is/40085/?COMMAND=DisplayBericht&FIS=203&OBJECT=40085&MODE=METADATA, Accessed 10 Feb 2015
11. Bendixen HJ (1999) Hygienic safety: results of scientific investigations in Denmark (sanitation requirements in Danish Biogas Plants). In: Böhm R, Wellinger A (eds) Proceedings of the IEA Workshop "Hygienic and environmental aspects of anaerobic digestion: Legislation and experiences in Europe". Universität Hohenheim, Stuttgart, Germany, pp 27–47

12. Lebuhn M, Wilderer P (2006) Biogastechnologie zur umweltverträglichen Flüssigmistver wertung und Energiegewinnung in Wasserschutzgebieten: wasserwirtschaftliche und hygienische Begleituntersuchung - Projektteil: Mikrobiologische, parasitologische und virologische Untersuchungen. Technische Universität München, Germany, pp 1–102, http://www.lfl.bayern.de/mam/cms07/publikationen/daten/schriftenreihe/p_23481.pdf. Accessed 10 Feb 2015
13. Clarke A, Rothery P (2008) Scaling of body temperature in mammals and birds. Funct Ecol 22:58–67. doi:10.1111/j.1365-2435.2007.01341.x
14. Leben C (1981) How plant pathogenic bacteria survive. Plant Dis 65:633–637
15. Arvizu-Gómez JL, Hernández-Morales A, Aguilar JRP, Álvarez-Morales A (2013) Transcriptional profile of *P. syringae* pv. *phaseolicola* NPS3121 at low temperature: physiology of phytopathogenic bacteria. BMC Microbiol 13:81
16. Smirnova A, Li H, Weingart H et al (2001) Thermoregulated expression of virulence factors in plant-associated bacteria. Arch Microbiol 176:393–399. doi:10.1007/s002030100344
17. Ratkowsky DA, Lowry RK, McMeekin TA et al (1983) Model for bacterial culture growth rate throughout the entire biokinetic temperature range. J Bacteriol 154:1222–1226
18. Huang L, Hwang A, Phillips J (2011) Effect of temperature on microbial growth rate-mathematical analysis: the Arrhenius and Eyring-Polanyi connections. J Food Sci 76: E553–E560. doi:10.1111/j.1750-3841.2011.02377.x
19. FNR (2013) Leitfaden Biogas—Von der Gewinnung zur Nutzung, 6th edn. Fachagentur Nachwachsende Rohstoffe, Gülzow
20. Oragui J (2003) Viruses in Faeces. In: Mara D, Horan N (eds) The handbook of water and wastewater microbiology. Academic Press Elsevier, London, pp 473–476
21. Yenigün O, Demirel B (2013) Ammonia inhibition in anaerobic digestion: a review. Process Biochem 48:901–911. doi:10.1016/j.procbio.2013.04.012
22. Andrade D, Marín Pérez C, Heuwinkel H, et al (2009) Biogasgewinnung aus Grassilage: Untersuchungen zur Prozessstabilität. In: Internationale Wissenschaftstagung Biogas Science 2009—Band 3. Freising, Germany, pp 529–538
23. Leitão RC, Vanhaandel A, Zeeman G, Lettinga G (2006) The effects of operational and environmental variations on anaerobic wastewater treatment systems: a review. Bioresour Technol 97:1105–1118. doi:10.1016/j.biortech.2004.12.007
24. Khalid A, Arshad M, Anjum M et al (2011) The anaerobic digestion of solid organic waste. Waste Manag 31:1737–1744. doi:10.1016/j.wasman.2011.03.021
25. Cioabla AE, Ionel I, Dumitrel G-A et al (2012) Comparative study on factors affecting anaerobic digestion of agricultural vegetal residues. Biotechnol Biofuels 5:1
26. Weiland P (2010) Biogas production: current state and perspectives. Appl Microbiol Biotechnol 85:849–860. doi:10.1007/s00253-009-2246-7
27. Schink B (2007) Wachstum und Ernährung der Mikroorganismen. In: Fuchs G (ed) Allgemeine Mikrobiologie, 8th edn. Georg Thieme Verlag, Leipzig, pp 155–192
28. Chen Y, Cheng JJ, Creamer KS (2008) Inhibition of anaerobic digestion process: a review. Bioresour Technol 99:4044–4064. doi:10.1016/j.biortech.2007.01.057
29. Ottoson JR, Schnürer A, Vinnerås B (2008) In situ ammonia production as a sanitation agent during anaerobic digestion at mesophilic temperature. Lett Appl Microbiol 46:325–330. doi:10.1111/j.1472-765X.2007.02317.x
30. Colleran E, Finnegan S, Lens P (1995) Anaerobic treatment of sulphate-containing waste streams. Antonie Van Leeuwenhoek 67:29–46. doi:10.1007/BF00872194
31. Henry DP, Frost AJ, Samuel Jl, et al (1983) Factors affecting the survival of *Salmonella* and *Escherichia coli* in anaerobically fermented pig waste. J Appl Bacteriol 55:89–95. doi:10.1111/j.1365-2672.1983.tb02651.x
32. Kunte Yeole, Chiplonkar Ranade (1998) Inactivation of *Salmonella typhi* by high levels of volatile fatty acids during anaerobic digestion. J Appl Microbiol 84:138–142. doi:10.1046/j.1365-2672.1997.00335.x

33. De Baere LA, Devocht M, Van Assche P, Verstraete W (1984) Influence of high NaCl and NH_4Cl salt levels on methanogenic associations. Water Res 18:543–548. doi:10.1016/0043-1354(84)90201-X
34. Soto M, Méndez R, Lema JM (1993) Sodium inhibition and sulphate reduction in the anaerobic treatment of mussel processing wastewaters. J Chem Technol Biotechnol 58:1–7
35. Salvador AF, Cavaleiro AJ, Sousa DZ et al (2013) Endurance of methanogenic archaea in anaerobic bioreactors treating oleate-based wastewater. Appl Microbiol Biotechnol 97:2211–2218. doi:10.1007/s00253-012-4061-9
36. Lin J, Smith MP, Chapin KC et al (1996) Mechanisms of acid resistance in enterohemorrhagic *Escherichia coli*. Appl Environ Microbiol 62:3094–3100
37. Duniere L, Gleizal A, Chaucheyras-Durand F et al (2011) Fate of *Escherichia coli* O26 in corn silage experimentally contaminated at ensiling, at silo opening, or after aerobic exposure, and protective effect of various bacterial inoculants. Appl Environ Microbiol 77:8696–8704. doi:10.1128/AEM.06320-11
38. Bandte M, Schleusner Y, Heiermann M et al (2013) Viability of plant-pathogenic fungi reduced by anaerobic digestion. BioEnergy Res 6:966–973. doi:10.1007/s12155-013-9326-3
39. Regulation (EC) No 1069/2009 (2013) Regulation (EC) No 1069/2009 of the European Parliament and of the Council of 21 October 2009 laying down health rules as regards animal by-products and derived products not intended for human consumption and repealing Regulation (EC) No 1774/2002 (Animal by-products Regulation), amended by Directive 2010/63/EU of the European Parliament and of the Council of 22 September 2010 and Council Regulation (EU) No 1385/2013 of 17 December 2013
40. Commission Regulation (EU) No 142/2011 (2012) Commission Regulation (EU) No 142/2011 of 25 February 2011 implementing Regulation (EC) No 1069/2009 of the European Parliament and of the Council laying down health rules as regards animal by-products and derived products not intended for human consumption and implementing Council Directive 97/78/EC as regards certain samples and items exempt from veterinary checks at the border under that Directive, amended by Commission Regulation (EU) No 749/2011 of 29 July 2011, Commission Regulation (EU) No 1063/2012 of 13 November 2012 and Commission Implementing Regulation (EU) No 1097/2012 of 23 November 2012
41. Plöchl M, Heiermann M, Rodemann B et al (2014) Kinetics of inactivation and dilution effects on the mass balance of fungal phytopathogens in anaerobic digesters. J Environ Manage 133:116–120. doi:10.1016/j.jenvman.2013.11.047
42. Reinhold G, Jahn O (2004) Hygienisierende Wirkungen der Biogaserzeugung auf die Gärsubstrate. In: Proceedings of the 116th VDLUFA congress, Rostock, Germany, 13–17 Sept 2004
43. Lebuhn M, Fröschle B (2012) Hygienische Aspekte beim Einsatz von Gärresten. In: Bayerische Landesanstalt für Landwirtschaft (ed) Düngung mit Biogasgärresten—effektiv—umweltfreundlich—bodenschonend—10. Kulturlandschaftstag, LfL-Schriftenreihe, Nr. 11, pp 59–71, http://www.lfl.bayern.de/mam/cms07/publikationen/daten/schriftenreihe/p_45175.pdf#page=59. Accessed 18 March 2015
44. Wang G, Zhao T, Doyle MP (1996) Fate of enterohemorrhagic *Escherichia coli* O157: H7 in bovine feces. Appl Environ Microbiol 62:2567–2570
45. Nicholson FA, Groves SJ, Chambers BJ (2005) Pathogen survival during livestock manure storage and following land application. Bioresour Technol 96:135–143. doi:10.1016/j.biortech.2004.02.030
46. Haumacher R, Schnaufer S, Philipp W, et al (2012) Abschlussbericht zum Forschungsvorhaben: Hygieneaspekte organischer Düngemittel und ihre Anwendung, p 315
47. Slepecky RA, Hemphill HE (2006) The genus *Bacillus*—nonmedical. In: Dworkin M, Falkow S, Rosenberg E et al (eds) The prokaryotes, 3rd edn. Springer, US, pp 530–562
48. Bagge E, Sahlström L, Albihn A (2005) The effect of hygienic treatment on the microbial flora of biowaste at biogas plants. Water Res 39:4879–4886. doi:10.1016/j.watres.2005.03.016

49. Bagge E (2009) Hygiene aspects of the biogas process with emphasis on spore-forming bacteria. Dissertation, Swedish University of Agricultural Sciences, Uppsala, Sweden
50. Tavares CRG, Benatti CT, Filho BPD (2002) Treatment of sewage sludge generated in municipal wastewater treatment plants. Appl Biochem Biotechnol 98–100:935–945. doi:10.1385/ABAB:98-100:1-9:935
51. Larsen HE, Munch B, Schlundt J (1994) Use of indicators for monitoring the reduction of pathogens in animal waste treated in biogas plants. Zentralblatt Hyg Umweltmed = Int J Hyg Environ Med 195:544–555
52. Watcharasukarn M, Kaparaju P, Steyer J-P et al (2009) Screening *Escherichia coli*, *Enterococcus faecalis*, and *Clostridium perfringens* as indicator organisms in evaluating pathogen-reducing capacity in biogas plants. Microb Ecol 58:221–230. doi:10.1007/s00248-009-9497-9
53. Aitken MD, Sobsey MD, Van Abel NA et al (2007) Inactivation of *Escherichia coli* O157: H7 during thermophilic anaerobic digestion of manure from dairy cattle. Water Res 41:1659–1666. doi:10.1016/j.watres.2007.01.034
54. Pandey PK, Soupir ML (2011) *Escherichia coli* inactivation kinetics in anaerobic digestion of dairy manure under moderate, mesophilic and thermophilic temperatures. AMB Express 1:18
55. RKI (2009) RKI-Ratgeber Infektionskrankheiten—Merkblätter für Ärzte—Salmonellose (Salmonellen-Gastroenteritis). Epidemiologisches Bulletin des Robert Koch-Instituts 13:117–126. http://www.rki.de/DE/Content/Infekt/EpidBull/Archiv/2009/13/Tabelle.html. Accessed 10 Feb 2015
56. Strauch D (1996) Hygieneaspekte bei der Cofermentation. In: Braun R (ed) Vorträge und Podiumsdiskussion im Rahmen der Fachtagung "Internationale Erfahrungen mit der Verwertung biogener Abfälle zur Biogasproduktion." Tulln,, Austria, pp 53–92. http://www.umweltbundesamt.at/fileadmin/site/publikationen/CP014.pdf#page=59. Accessed 19 Dec 2014
57. BioAbfV (2013) Bioabfallverordnung in der Fassung der Bekanntmachung vom 4. April 2013 (BGBl. I S. 658), die zuletzt durch Artikel 5 der Verordnung vom 5. Dezember 2013 (BGBl. I S. 4043) geändert worden ist, pp 1–57
58. TierNebV (2012) Tierische Nebenprodukte-Beseitigungsverordnung vom 27. Juli 2006 (BGBl. I S. 1735), die zuletzt durch Artikel 2 der Verordnung vom 23. April 2012 (BGBl. I S. 611) geändert worden ist, pp 1–21
59. DüMV (2012) Düngemittelverordnung vom 5. Dezember 2012 (BGBl. I S. 2482), pp 1–113
60. Smith SR, Lang NL, Cheung KHM, Spanoudaki K (2005) Factors controlling pathogen destruction during anaerobic digestion of biowastes. Waste Manag 25:417–425. doi:10.1016/j.wasman.2005.02.010
61. Franke-Whittle IH, Insam H (2013) Treatment alternatives of slaughterhouse wastes, and their effect on the inactivation of different pathogens: a review. Crit Rev Microbiol 39:139–151. doi:10.3109/1040841X.2012.694410
62. Paavola T, Syväsalo E, Rintala J (2006) Co-digestion of manure and biowaste according to the EC animal by-products regulation and Finnish national regulations. Water Sci Technol 53:223. doi:10.2166/wst.2006.253
63. Paavola T, Rintala J (2008) Effects of storage on characteristics and hygienic quality of digestates from four co-digestion concepts of manure and biowaste. Bioresour Technol 99:7041–7050. doi:10.1016/j.biortech.2008.01.005
64. RKI (2005) Campylobacter-Infektionen—RKI-Ratgeber für Ärzte. http://www.rki.de/DE/Content/Infekt/EpidBull/Merkblaetter/Ratgeber_Campylobacter.html. Accessed 10 Feb 2015
65. RKI (2006) Ausgewählte Zoonosen im Jahr 2005: Durch Lebensmittel übertragbare bakterielle gastrointestinale Infektionen. Epidemiologisches Bulletin des Robert Koch-Instituts 41:351–362. http://www.rki.de/DE/Content/Infekt/EpidBull/Archiv/2006/Ausgabenlinks/41_06.html. Accessed 10 Feb 2015

66. Martens W, Böhm R (2009) Overview of the ability of different treatment methods for liquid and solid manure to inactivate pathogens. Bioresour Technol 100:5374–5378. doi:10.1016/j.biortech.2009.01.014
67. Millner PD (2009) Manure management. In: Sapers GM, Solomon EB, Matthews KR (eds) The produce contamination problem—causes and solutions. Academic Press, San Diego, pp 79–104
68. Hutchison ML, Walters LD, Avery SM et al (2005) Analyses of livestock production, waste storage, and pathogen levels and prevalences in farm manures. Appl Environ Microbiol 71:1231–1236. doi:10.1128/AEM.71.3.1231-1236.2005
69. Sahlström L, Bagge E, Emmoth E et al (2008) A laboratory study of survival of selected microorganisms after heat treatment of biowaste used in biogas plants. Bioresour Technol 99:7859–7865. doi:10.1016/j.biortech.2007.09.071
70. Stampi S, De Luca G, Varoli O, Zanetti F (1999) Occurrence, removal and seasonal variation of thermophilic campylobacters and *Arcobacter* in sewage sludge. ZBL Hyg Umweltmed 202:19–27. doi:10.1016/S0934-8859(99)80048-0
71. Godfree A, Farrell J (2005) Processes for managing pathogens. J Environ Qual 34:105–113
72. Lebuhn M, Effenberger M, Garcés G et al (2005) Hygienization by anaerobic digestion: comparison between evaluation by cultivation and quantitative real-time PCR. Water Sci Technol 52:93–99
73. Wagner AO, Gstraunthaler G, Illmer P (2008) Survival of bacterial pathogens during the thermophilic anaerobic digestion of biowaste: laboratory experiments and in situ validation. Anaerobe 14:181–183. doi:10.1016/j.anaerobe.2008.03.004
74. Wagner AO, Malin C, Gstraunthaler G, Illmer P (2009) Survival of selected pathogens in diluted sludge of a thermophilic waste treatment plant and in NaCl-solution under aerobic and anaerobic conditions. Waste Manag 29:425–429. doi:10.1016/j.wasman.2008.03.003
75. Ade-Kappelmann K (2008) Untersuchungen zur seuchenhygienischen Unbedenklichkeit von Gärresten aus Bioabfällen nach der Behandlung in Anaerobanlagen. Inaugural dissertation, Freie Universität Berlin, Berlin, Germany
76. Massé D, Gilbert Y, Topp E (2011) Pathogen removal in farm-scale psychrophilic anaerobic digesters processing swine manure. Bioresour Technol 102:641–646. doi:10.1016/j.biortech.2010.08.020
77. Manyi-Loh CE, Mamphweli SN, Meyer EL et al (2014) Inactivation of selected bacterial pathogens in dairy cattle manure by mesophilic anaerobic digestion (balloon type digester). Int J Environ Res Public Health 11:7184–7194. doi:10.3390/ijerph110707184
78. Horan NJ, Fletcher L, Betmal SM et al (2004) Die-off of enteric bacterial pathogens during mesophilic anaerobic digestion. Water Res 38:1113–1120. doi:10.1016/j.watres.2003.12.004
79. Kearney TE, Larkin MJ, Frost JP, Levett PN (1993) Survival of pathogenic bacteria during mesophilic anaerobic digestion of animal waste. J Appl Bacteriol 75:215–219. doi:10.1111/j.1365-2672.1993.tb02768.x
80. Kearney TE, Larkin MJ, Levett PN (1994) Metabolic activity of pathogenic bacteria during semicontinuous anaerobic digestion. Appl Environ Microbiol 60:3647–3652
81. Oliver JD (2005) The viable but nonculturable state in bacteria. J Microbiol 43:93–100
82. Murphy C, Carroll C, Jordan KN (2006) Environmental survival mechanisms of the foodborne pathogen *Campylobacter jejuni*. J Appl Microbiol 100:623–632. doi:10.1111/j.1365-2672.2006.02903.x
83. Elving J (2009) Pathogen inactivation and regrowth in organic waste during biological treatment. Dissertation, Swedish University of Agricultural Sciences, Uppsala, Sweden
84. USFDA (2000) Safe practices for food processes—kinetics of microbial inactivation for alternative food processing technologies—overarching principles: kinetics and pathogens of concern for all technologies. http://www.fda.gov/Food/FoodScienceResearch/SafePracticesforFoodProcesses/ucm100198.htm. Accessed 21 Jan 2015
85. Kaper JB, Nataro JP, Mobley HLT (2004) Pathogenic *Escherichia coli*. Nat Rev Microbiol 2:123–140. doi:10.1038/nrmicro818
86. Nataro JP, Kaper JB (1998) Diarrheagenic *Escherichia coli*. Clin Microbiol Rev 11:142–201

87. Ammon A, Schmidt K, Bräunig J (2000) Lebensmittelinfektionen in Deutschland. Bundesgesundheitsblatt-Gesundheitsforschung-Gesundheitsschutz 43:751–757
88. Bülte M, Montenegro MA, Helmuth R et al (1990) Nachweis von Verotoxin-bildenden *E. coli* (VTEC) bei gesunden Rindern und Schweinen mit dem DNS-DNS-Koloniehybridisierungsverfahren. Berl Münch Tierärztl Wochenschr 103:380–384
89. Beutin L, Geier D, Steinrück H et al (1993) Prevalence and some properties of verotoxin (Shiga-like toxin)-producing *Escherichia coli* in seven different species of healthy domestic animals. J Clin Microbiol 31:2483–2488
90. Baljer G, Wieler L (1999) Animals as a source of infections for humans—diseases caused by EHEC. DTW Dtsch Tierarztl Wochenschr 106:339–343
91. Messelhäusser U, Fröschle B, Lebuhn M, Höller C (2015) Detection of Shigatoxin-producing *Escherichia coli* (STEC) and Enteropathogenic *Escherichia coli* (EPEC) in biogas plants in combination with germ carrier experiments (submitted)
92. RKI (ed) (2011) Erkrankungen durch Enterohämorrhagische *Escherichia coli* (EHEC)—RKI Ratgeber für Ärzte. Epidemiologisches Bulletin des Robert Koch-Instituts 31/1999
93. Breves G (2012) Abschlussbericht—Einschätzung des Risikopotenzials der Biogastechnologie in mikrobiologischer Hinsicht. 1–46, Stiftung Tierärztliche Hochschule Hannover, Hannover, Germany, http://www.ml.niedersachsen.de/download/72762. Accessed 09 Feb 2015
94. Bonetta S, Ferretti E, Bonetta S et al (2011) Microbiological contamination of digested products from anaerobic co-digestion of bovine manure and agricultural by-products: Microbial contamination of digestate. Lett Appl Microbiol 53:552–557. doi:10.1111/j.1472-765X.2011.03148.x
95. Spinks AT, Dunstan RH, Harrison T et al (2006) Thermal inactivation of water-borne pathogenic and indicator bacteria at sub-boiling temperatures. Water Res 40:1326–1332. doi:10.1016/j.watres.2006.01.032
96. Doyle MP, Schoeni JL (1984) Survival and growth characteristics of *Escherichia coli* associated with hemorrhagic colitis. Appl Environ Microbiol 48:855–856
97. RKI (2011) Steckbriefe seltener Infektionskrankheiten. http://www.rki.de/DE/Content/InfAZ/Steckbriefe/Steckbriefe_120606.html. Accessed 10 Feb 2015
98. Percival SL, Williams DW (2014) *Yersinia*. In: Percival SL, Yates MV, Williams DW, et al (eds) Microbiology of waterborne diseases, 2nd edn. Academic Press Elsevier, San Diego, pp 249–260
99. Guan TY, Holley RA (2003) Pathogen survival in swine manure environments and transmission of human enteric illness—a review. J Environ Qual 32:383–392
100. Hartung M, Gerigk K (1991) *Yersinia* in effluents from the food-processing industry. Rev Sci Tech IOE 10:799–811
101. Bhaduri S (2002) Comparison of multiplex PCR, PCR-ELISA and fluorogenic 5′ nuclease PCR assays for detection of plasmid-bearing virulent *Yersinia enterocolitica* in swine feces. Mol Cell Probes 16:191–196. doi:10.1006/mcpr.2002.0408
102. Burtscher C, Wuertz S (2003) Evaluation of the use of PCR and reverse transcriptase PCR for detection of pathogenic bacteria in biosolids from anaerobic digestors and aerobic composters. Appl Environ Microbiol 69:4618–4627. doi:10.1128/AEM.69.8.4618-4627.2003
103. Côté C, Massé DI, Quessy S (2006) Reduction of indicator and pathogenic microorganisms by psychrophilic anaerobic digestion in swine slurries. Bioresour Technol 97:686–691. doi:10.1016/j.biortech.2005.03.024
104. Bibby K, Viau E, Peccia J (2010) Pyrosequencing of the 16S rRNA gene to reveal bacterial pathogen diversity in biosolids. Water Res 44:4252–4260. doi:10.1016/j.watres.2010.05.039
105. Drča M (2007) Seuchenhygienisch-mikrobiologische Untersuchungen an einer mesophil betriebenen Biogasanlage zur Verwertung von Speiseresten in Verbindung mit methodischen Untersuchungen zum Nachweis von Salmonellen und *Escherichia coli* aus biologischem Material. Inaugural dissertation, Universität Leipzig, Leipzig, Germany

106. Kearney TE, Larkin MJ, Levett PN (1993) The effect of slurry storage and anaerobic digestion on survival of pathogenic bacteria. J Appl Bacteriol 74:86–93. doi:10.1111/j.1365-2672.1993.tb03000.x
107. Avery LM, Anchang KY, Tumwesige V et al (2014) Potential for pathogen reduction in anaerobic digestion and biogas generation in Sub-Saharan Africa. Biomass Bioenergy 70:112–124. doi:10.1016/j.biombioe.2014.01.053
108. Smith NH, Gordon SV, de la Rua-Domenech R et al (2006) Bottlenecks and broomsticks: the molecular evolution of *Mycobacterium bovis*. Nat Rev Microbiol 4:670–681. doi:10.1038/nrmicro1472
109. Büttner M, Gerbermann H, Naumann L, et al (2006) Paratuberkulose beim Rind—Morbus Crohn beim Menschen: ein ursächlicher Zusammenhang? Bayerisches Landesamt für Gesundheit und Lebensmittelsicherheit (ed), pp 1–114
110. Leite FL, Reinhardt TA, Bannantine JP, Stabel JR (2015) Envelope protein complexes of *Mycobacterium avium* subsp. *paratuberculosis* and their antigenicity. Vet Microbiol 175:275–285. doi:10.1016/j.vetmic.2014.11.009
111. Sartor RB (2006) Mechanisms of disease: pathogenesis of Crohn's disease and ulcerative colitis. Nat Clin Pract Gastroenterol Hepatol 3:390–407. doi:10.1038/ncpgasthep0528
112. Percival SL, Williams DW (2014) *Mycobacterium*. In: Percival SL, Yates MV, Williams DW, et al (eds) Microbiology of waterborne diseases, 2nd edn. Academic Press Elsevier, San Diego, pp 177–208
113. Wright PE, Inglis SF, Stehman SM, Bonhotal J (2003) Reduction of selected pathogens in anaerobic digestion. In: Proceedings of the 9th international symposium of animal, agricultural and food processing wastes, pp 11–14
114. Slana I, Pribylova R, Kralova A, Pavlik I (2011) Persistence of *Mycobacterium avium* subsp. *paratuberculosis* at a farm-scale biogas plant supplied with manure from paratuberculosis-affected dairy cattle. Appl Environ Microbiol 77:3115–3119. doi:10.1128/AEM.02407-10
115. Olsen JE, Jørgensen JB, Nansen P (1985) On the reduction of *Mycobacterium paratuberculosis* in bovine slurry subjected to batch mesophilic or thermophilic anaerobic digestion. Agric Wastes 13:273–280. doi:10.1016/0141-4607(85)90052-6
116. Bendixen H (1994) Safeguards against pathogens in Danish biogas plants. Water Sci Technol 30:171–180
117. Krause L, Diaz NN, Edwards RA et al (2008) Taxonomic composition and gene content of a methane-producing microbial community isolated from a biogas reactor. J Biotechnol 136:91–101. doi:10.1016/j.jbiotec.2008.06.003
118. Kumar G (2015) Challenges in developing biotoxin inhibitors. In: Gopalakrishnakone P, Balali-Mood M, Llewellyn L, Singh BR (eds) Biological toxins and bioterrorism. Springer, Dordrecht, pp 357–373
119. Collins Md, East AK (1998) Phylogeny and taxonomy of the food-borne pathogen *Clostridium botulinum* and its neurotoxins. J Appl Microbiol 84:5–17
120. Parkinson NG, Ito KA (2006) *Clostridium botulinum*. In: Riemann H, Cliver DO (eds) Foodborne infections and intoxications. Elsevier Academic Press, Amsterdam, pp 485–522
121. Smith LD (1978) The occurrence of *Clostridium botulinum* and *Clostridium tetani* in the soil of the United States. Health Lab Sci 15:74–80
122. Notermans SH, Havelaar AH, Dufrenne JB, Oosterom J (1985) Incidence of *Clostridium botulinum* on cattle farms. Tijdschr Diergeneeskd 110:175–180
123. Smart JL, Jones TO, Clegg FG, McMurtry MJ (1987) Poultry waste associated type C botulism in cattle. Epidemiol Infect 98:73–79
124. McLoughlin M, Mcllroy S, Neill S (1988) A major outbreak of botulism in cattle being fed ensiled poultry litter. Vet Rec 122:579–581. doi:10.1136/vr.122.24.579
125. Klarmann D (1989) The detection of *Clostridium botulinum* in fecal samples of cattle and swine and in the raw material and animal meal of different animal body rendering plants. Berl Münch Tierärztl Wochenschr 102:84–86

126. Ortolani EL, Brito LA, Mori CS et al (1997) Botulism outbreak associated with poultry litter consumption in three Brazilian cattle herds. Vet Hum Toxicol 39:89–92
127. Dahlenborg M, Borch E, Rådström P (2001) Development of a combined selection and enrichment PCR procedure for *Clostridium botulinum* types B, E, and F and its use to determine prevalence in fecal samples from slaughtered pigs. Appl Environ Microbiol 67:4781–4788. doi:10.1128/AEM.67.10.4781-4788.2001
128. Dahlenborg M, Borch E, Rådström P (2003) Prevalence of *Clostridium botulinum* types B, E and F in faecal samples from Swedish cattle. Int J Food Microbiol 82:105–110
129. Driehuis F (2013) Silage and the safety and quality of dairy foods: a review. Agric Food Sci 22:16–34
130. Tanner FW, Dack GM (1922) *Clostridium botulinum*. J Infect Dis 31:92–100
131. Bagge E, Persson M, Johansson K-E (2010) Diversity of spore-forming bacteria in cattle manure, slaughterhouse waste and samples from biogas plants: Spore-forming bacteria in biogas plants. J Appl Microbiol 109:1549–1565. doi:10.1111/j.1365-2672.2010.04790.x
132. Fröschle B, Messelhäusser U, Höller C, Lebuhn M (2015) Fate of *Clostridium botulinum* and incidence of pathogenic clostridia in biogas processes. J Appl Microbiol (Accepted for publication)
133. Myllykoski J, Lindström M, Keto-Timonen R et al. (2009) Type C bovine botulism outbreak due to carcass contaminated non-acidified silage. Epidemiol Infect 137:284–293. doi:10.1017/S0950268808000939
134. Tebbe C, Dohrmann AB, Baumert S, et al (2007) Untersuchungen zum qualitativen und quantitativen Vorkommen von *Clostridium botulinum* in Substraten und Gärrückständen von Biogasanlagen. Bundesforschungsanstalt für Landwirtschaft (ed) Braunschweig, Germany, pp 1–112
135. Breves G (2011) Untersuchungen zum mikrobiologischen Risikopotenzial von Gärsubstraten und Gärresten aus niedersächsischen Biogasanlagen. Stiftung Tierärztliche Hochschule Hannover, Hannover, Germany, http://www.ml.niedersachsen.de/portal/live.php?navigation_id=1375&article_id=101070&_psmand=7. Accessed 09 Feb 2015
136. Dohrmann AB, Baumert S, Klingebiel L et al (2011) Bacterial community structure in experimental methanogenic bioreactors and search for pathogenic clostridia as community members. Appl Microbiol Biotechnol 89:1991–2004. doi:10.1007/s00253-010-2955-y
137. Eikmeyer FG, Rademacher A, Hanreich A et al (2013) Detailed analysis of metagenome datasets obtained from biogas-producing microbial communities residing in biogas reactors does not indicate the presence of putative pathogenic microorganisms. Biotechnol Biofuels 6:49
138. Köhler B (2011) Wechselwirkungen von Clostridien-Toxininfektionen mit der Umwelt unter besonderer Berücksichtigung von *Clostridium botulinum* und Entsorgungsverfahren in der Tierhaltung. Tierärztliche Hochschule Hannover, Hannover, Germany. http://www.ml.niedersachsen.de/portal/live.php?navigation_id=1375&article_id=101070&_psmand=7. Accessed 10 Feb 2015
139. Köhler B (2012) Hygienische Probleme von Biogasanlagen unter besonderer Berücksichtigung von *Clostridium botulinum* und anderer pathogener Clostridien. Jena, Germany. http://www.tll.de/imperia/ft_biog/tbi60412.pdf. Accessed 10 Feb 2015
140. Hatheway CL (1990) Toxigenic clostridia. Clin Microbiol Rev 3:66–98. doi:10.1128/CMR.3.1.66
141. Breves G (2014) Abschlussbericht—Mikrobiologisches Risikopotenzial von Biogasanlagen unter besonderer Berücksichtigung von Hühnertrockenkot als Gärsubstrat (Teilprojekt 2), pp 1–114. Stiftung Tierärztliche Hochschule Hannover, Hannover, Germany. http://www.bmel.de/SharedDocs/Downloads/Tier/Tiergesundheit/Tierseuchen/Botulismus-Abschlussbericht-Hannover.html. Accessed 09 Feb 2015
142. Neuhaus J, Shehata AA, Krüger M (2014) Detection of pathogenic clostridia in biogas plant wastes. Folia Microbiol (Praha). doi:10.1007/s12223-014-0334-2
143. Aitken MD, Sobsey MD, Shehee M et al (2005) Laboratory evaluation of thermophilic-anaerobic digestion to produce class A biosolids. 2. Inactivation of pathogens

and indicator organisms in a continuous-flow reactor followed by batch treatment. Water Environ Res 77:3028–3036. doi:10.2175/106143005X73910
144. Huong LQ, Forslund A, Madsen H, Dalsgaard A (2014) Survival of *Salmonella* spp. and fecal indicator bacteria in Vietnamese biogas digesters receiving pig slurry. Int J Hyg Environ Health 217:785–795. doi:10.1016/j.ijheh.2014.04.004
145. Songer JG (1998) Clostridial diseases of small ruminants. Vet Res 29:219–232
146. RKI (2009) RKI-Ratgeber Infektionskrankheiten—Merkblätter für Ärzte—*Clostridium difficile*. Epidemiol Bull Robert Koch-Inst 24:233–239. http://www.rki.de/DE/Content/Infekt/EpidBull/Archiv/2009/24/Tabelle.html. Accessed 10 Feb 2015
147. Bagge E, Lewerin SS, Johansson K-E (2009) Detection and identification by PCR of *Clostridium chauvoei* in clinical isolates, bovine faeces and substrates from biogas plant. Acta Vet Scand 51:8. doi:10.1186/1751-0147-51-8
148. Viau E, Peccia J (2009) Survey of Wastewater indicators and human pathogen genomes in biosolids produced by class A and class B stabilization treatments. Appl Environ Microbiol 75:164–174. doi:10.1128/AEM.01331-08
149. Wirth R, Kovács E, Maróti G et al (2012) Characterization of a biogas-producing microbial community by short-read next generation DNA sequencing. Biotechnol Biofuels 5:41
150. Ammon A, Bräunig J (2002) Lebensmittelbedingte Erkrankungen in Deutschland. In: RKI (ed) Gesundheitsberichterstattung des Bundes Heft 01/02. Berlin, Germany, pp 1–12
151. Rasko DA, Altherr MR, Han CS, Ravel J (2005) Genomics of the *Bacillus cereus* group of organisms. FEMS Microbiol Rev 29:303–329. doi:10.1016/j.fmrre.2004.12.005
152. Vissers MMM, Te Giffel MC, Driehuis F et al (2007) Minimizing the level of *Bacillus cereus* spores in farm tank milk. J Dairy Sci 90:3286–3293. doi:10.3168/jds.2006-873
153. Lebuhn M, Effenberger M, Garces G et al (2004) Evaluating real-time PCR for the quantification of distinct pathogens and indicator organisms in environmental samples. Water Sci Technol 50:263–270
154. Govasmark E, Stäb J, Holen B et al (2011) Chemical and microbiological hazards associated with recycling of anaerobic digested residue intended for agricultural use. Waste Manag 31:2577–2583. doi:10.1016/j.wasman.2011.07.025
155. Guy RA, Payment P, Krull UJ, Horgen PA (2003) Real-time PCR for quantification of *Giardia* and *Cryptosporidium* in environmental water samples and sewage. Appl Environ Microbiol 69:5178–5185. doi:10.1128/AEM.69.9.5178-5185.2003
156. Hill A, Nally P, Chalmers RM et al (2011) Quantitative risk assessment for zoonotic transmission of *Cryptosporidium parvum* infection attributable to recreational use of farmland. Zoonoses Public Health 58:323–333. doi:10.1111/j.1863-2378.2010.01350.x
157. Robertson LJ (2015) Waterborne zoonoses: *Cryptosporidium* and Cryptosporidiosis: a small parasite that makes a big splash. In: Sing A (ed) zoonoses—infections affecting humans and animals. Springer, Dordrecht, pp 803–819
158. Hutchison ML, Walters LD, Avery SM et al (2004) Levels of zoonotic agents in British livestock manures. Lett Appl Microbiol 39:207–214. doi:10.1111/j.1472-765X.2004.01564.x
159. Garcés G, Effenberger M, Najdrowski M et al (2006) Quantification of *Cryptosporidium parvum* in anaerobic digesters treating manure by (reverse-transcription) quantitative real-time PCR, infectivity and excystation tests. Water Sci Technol 53:195. doi:10.2166/wst.2006.250
160. Chauret C, Springthorpe S, Sattar S (1999) Fate of *Cryptosporidium* oocysts, *Giardia* cysts, and microbial indicators during wastewater treatment and anaerobic sludge digestion. Can J Microbiol 45:257–262. doi:10.1139/w99-001
161. EFSA (2007) Opinion of the Scientific Panel on Biological Hazards on a request from the European Commission on the safety vis-à-vis biological risk of the mesophilic process of biogas and compost treatment of animal by-products (ABPs). EFSA J 465:1–16
162. Ongerth JE, Johnson RL, Macdonald SC et al (1989) Back-country water treatment to prevent giardiasis. Am J Public Health 79:1633–1637. doi:10.2105/AJPH.79.12.1633
163. Fayer R (1994) Effect of high temperature on infectivity of *Cryptosporidium parvum* oocysts in water. Appl Environ Microbiol 60:2732–2735

164. Plym-Forshell L (1995) Survival of salmonellas and *Ascaris suum* eggs in a thermophilic biogas plant. Acta Vet Scand 36:79–85
165. Rolle M, Mayr A (eds) (1993) Kapitel II Allgemeine Virologie - 2 Widerstandsfähigkeit, Inaktivierung, Desinfektion. In: Medizinische Mikrobiologie, Infektions- und Seuchenlehre für Tierärzte, Biologen, Agrarwissenschaftler und Interessierte aus benachbarten Fachgebieten: Lehrbuch für Praxis und Studium, 6th edn. Enke, Stuttgart, Germany, pp 163–167
166. Lund B, Jensen VF, Have P, Ahring B (1996) Inactivation of virus during anaerobic digestion of manure in laboratory scale biogas reactors. Antonie Van Leeuwenhoek 69:25–31. doi:10.1007/BF00641608
167. Gao T, Tong Y, Cao M et al (2013) Evaluation of the inactivation of human *Coxsackievirus* by thermophilic and mesophilic anaerobic digestion using integrated cell culture and reverse transcription real-time quantitative PCR. Water Res 47:4259–4264. doi:10.1016/j.watres.2013.04.046
168. Prado T, Gaspar AMC, Miagostovich MP (2014) Detection of enteric viruses in activated sludge by feasible concentration methods. Braz J Microbiol 45:343–349. doi:10.1590/S1517-83822014000100049
169. Bandte M, Pietsch M, Schultheiß U, et al (2012) Ein Verbundprojekt zum phytosanitären Risiko bei der anaeroben Vergärung von pflanzlichen Biomassen in Biogasanlagen. [A joint project on the phytosanitary risk associated with the anaerobic digestion of plant material in biogas plants]. Kurzfassungen der 58. Deutschen Pflanzenschutztagung "Pflanzenschutz – alternativlos". Julius-Kühn-Archiv, Braunschweig, Germany, pp 1–59
170. Rodemann B, Pottberg U, Pietsch M (2012) Untersuchungen zur Inaktivierung von Getreide- und Maispathogenen in Biogasanlagen. In: Kuratorium für Technik und Bauwesen in der Landwirtschaft e.V. (KTBL) (ed) Untersuchungen zum phytosanitären Risiko bei der anaeroben Vergärung von pflanzlichen Biomassen in Biogasanlagen. KTBL-Fachgespräch 14. 11. 2011. Berlin, Darmstadt, Germany, pp 17–26
171. Seigner L, Friedrich R, Kaemmerer D, et al (2010) Hygienisierungspotenzial des Biogasprozesses—Ergebnisse aus einem Forschungsprojekt. In: Bayerische Landesanstalt für Landwirtschaft (ed) LfL-Schriftenreihe, http://www.lfl.bayern.de/mam/cms07/publikationen/daten/schriftenreihe/p_40223.pdf. Accessed 10 Feb 2015
172. Termorshuizen AJ, Volker D, Blok WJ et al (2003) Survival of human and plant pathogens during anaerobic mesophilic digestion of vegetable, fruit, and garden waste. Eur J Soil Biol 39:165–171. doi:10.1016/S1164-5563(03)00032-3
173. Herrmann C, Heiermann M, Idler C, Prochnow A (2012) Particle size reduction during harvesting of crop feedstock for biogas production I: effects on ensiling process and methane yields. BioEnergy Res 5:926–936. doi:10.1007/s12155-012-9206-2
174. Westerman PR, Heiermann M, Pottberg U et al (2012) Weed seed survival during mesophilic anaerobic digestion in biogas plants. Weed Res 52:307–316. doi:10.1111/j.1365-3180.2012.00927.x
175. Westerman PR, Hildebrandt F, Gerowitt B (2012) Weed seed survival following ensiling and mesophilic anaerobic digestion in batch reactors. Weed Res 52:286–295. doi:10.1111/j.1365-3180.2012.00918.x
176. Westerman PR, Gerowitt B (2013) Weed seed survival during anaerobic digestion in biogas plants. Bot Rev 79:281–316. doi:10.1007/s12229-013-9118-7
177. Sölter U, Verschwele A, Starfinger U (2014) Das EU Projekt HALT Ambrosia—Fragen und Antworten. Julius-Kühn-Archiv Nr. 445, pp 156–160
178. Heiermann M, Herrmann C, Idler C, Starfinger U (2010) Can *Ambrosia* seeds survive the biogas process. In: Kollmann J, Van Mölken T, Raven HP (eds) Book Abstr. 6th NEOBIOTA conference biological invasions in a changing world from science to management, Copenhagen, Denmark, pp 1–256

Direct Interspecies Electron Transfer in Anaerobic Digestion: A Review

Charles-David Dubé and Serge R. Guiot

Abstract Direct interspecies electrons transfer (DIET) is a syntrophic metabolism in which free electrons flow from one cell to another without being shuttled by reduced molecules such as molecular hydrogen or formate. As more and more microorganisms show a capacity for electron exchange, either to export or import them, it becomes obvious that DIET is a syntrophic metabolism that is much more present in nature than previously thought. This article reviews literature related to DIET, specifically in reference to anaerobic digestion. Anaerobic granular sludge, a biofilm, is a specialized microenvironment where syntrophic bacterial and archaeal organisms grow together in close proximity. Exoelectrogenic bacteria degrading organic substrates or intermediates need an electron sink and electrotrophic methanogens represent perfect partners to assimilate those electrons and produce methane. The granule extracellular polymeric substances by making the biofilm matrix more conductive, play a role as electrons carrier in DIET.

Keywords Anaerobic digestion · Direct interspecies electron transfer (DIET) · Granular sludge · Extracellular polymeric substance (EPS) · Exoelectrogenic bacteria · Electrotrophic methanogen

---

NRC paper No. 55668

C.-D. Dubé
Department of Microbiology, Infectiology and Immunology, Université de Montréal, Montreal, Canada
e-mail: charles-david.dube@cnrc-nrc.gc.ca

C.-D. Dubé · S.R. Guiot (✉)
Bioengineering Group, Energy Mining and Environment, National Research Council Canada, Montreal, Canada
e-mail: serge.guiot@cnrc-nrc.gc.ca

G.M. Guebitz et al. (eds.), *Biogas Science and Technology*,
Advances in Biochemical Engineering/Biotechnology 151,
DOI 10.1007/978-3-319-21993-6_4

Contents

1 Introduction

The first scientific paper on microbial electricity generation was published in 1911 by Michael C. Potter of the University of Durham (UK) [1]. But it is only in the past few years that this capability became more than an academic curiosity. The reasons for this recent interest in using bacteria to generate electricity are a combination of the need for new sources of energy, discoveries about microbial physiology related to electron transport, and the advancement of fuel-cell technologies.

In a microbial fuel cell (MFC), bacteria are separated from a terminal electron acceptor at the cathode so that the only means for respiration is to transfer electrons to the anode [2, 3]. However, it is unlikely that the electricity generation has ever been a selective advantage during evolution, unlike the electron transfer to a natural extracellular acceptor. Exoelectrogenic microbes developed during evolution pathways to export electrons through the plasmic membrane to reduce external molecules, such as metal oxides (iron oxide (Fe_2O_3) or manganese oxide (IV), MnO_2) and humic substances, as a respiration mechanism for thereby acquiring energy and reproducing. For instance, bacteria of the genus *Geobacter* reduce Fe_2O_3 into soluble ferrous oxide (FeO). Fe_2O_3 is, as the anode of a MFC, an extracellular insoluble acceptor, thus requiring mechanisms for extracellular electron transfer [4]. As scientists start to exploit this capability to develop MFCs to produce renewable and alternative energies, it started to become obvious that the microbes in a natural environment could also benefit of these extracellular electron transfer pathways for optimizing their syntrophic associations. We review here literature reports that support the reality of the direct interspecies electron transfer (DIET), specifically in the anaerobic granular sludge used in industrial anaerobic digesters for wastewater biotreatment. The anaerobic granule is a stable, efficient and structured micro-environment where direct electron exchange between members of such consortium is likely to exist.

2 Likelihood of DIET in Anaerobic Granular Sludge

"What appears to be always and everywhere present [in life] is a continuous and rapid flow of electrons and protons passing through each and every living organism" [5]. This is presumably universal for each individual organism, but it is even of greater significance for syntrophic organisms. In methanogenic syntrophic communities, electrons and protons or reducing equivalents flow from one organism to the other by shuttle components such as molecular hydrogen. In the case of obligate syntrophy, the hydrogen-utilizing methanogens are essential to maintain the low concentrations of hydrogen that make the hydrogen-producing reaction sufficiently exergonic to support energy conservation (phosphorylation of ADP to ATP), hence cell maintenance and growth [6]. Obligate syntrophy was first recognized by Bryant and coworkers in 1967 [7] when they discovered that the originally believed pure culture *Methanobacillus omelianskii* was a co-culture of an ethanol-oxidizing bacterium and a methanogen, the latter scavenging the hydrogen produced by the former. Nowadays, it is well established that not only hydrogen but formate also plays a key role for electron transfer in methanogenic environments [8–11]. Optimization of large-scale anaerobic wastewater treatment relies on highly efficient methanogenic micro-ecosystems such as granular sludge, which develops in upflow anaerobic sludge blanket (UASB) reactors [12, 13]. It has been shown that aggregation of cells such as in methanogenic granules is a key factor for efficient methanization as a direct result of an efficient electron transfer between obligate H_2-producing acetogens (OHPA) and methanogens, especially as granules often present a layered architecture (Fig. 1) which provides niches for syntrophic associations and promotes the physical proximity between those syntrophs [14, 15].

Methane is not always the ultimate reduced product in anaerobic environments, and sulphur, metal oxides and humic substances are also used as final electron acceptors. Some of those molecules are kept outside of living cells and microorganisms developed specialized pathways to export electrons through their membrane to regenerate reduced coenzymes such as NADH or ferredoxin. For instance, *Geobacter sulfurreducens* uses a chain of trans-membrane cytochromes to transport electrons to acceptors outside the cell [16]. This property of dissimilatory reduction, now known to be widespread in the phylogenetic tree of bacteria, is used in MFCs and other bioelectrochemical systems (BES), where a positively charged electrode acts as a permanent electron acceptor.

In 2006, Stams and co-workers [11], then followed by others [17–21], suggested that a direct electron transfer excluding hydrogen and formate could happen between OHPA bacteria and methanogenic archaea in some environments. Incidentally, DIET might have been an explanation of previous observations that granular sludge was able to degrade propionate even though the Gibbs free energy change based on the dissolved H_2 measured in the liquid surrounding the granules was positive [22–24]. This emerging concept of DIET is surveyed here in the particular context of anaerobic granular sludge, an engineered methanogenic micro-ecosystem where the proximity and diversity of cell populations would allow for this exchange.

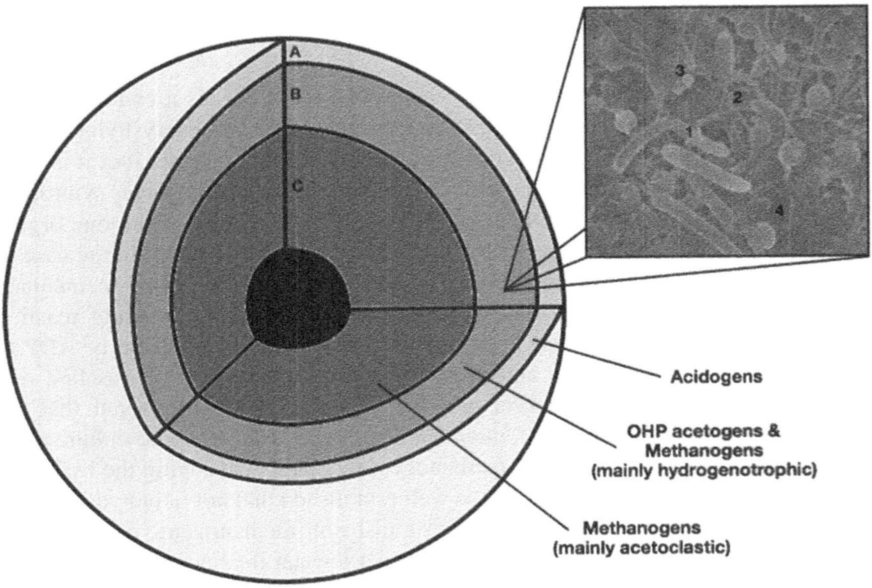

Fig. 1 Multilayered structure of the carbohydrate-fed anaerobic granule [15]. The syntrophic associations between obligate H_2-producing (OHP) acetogens and H_2-consuming microorganisms are prevailing in the middle layer (B) [30], where presumably direct interspecies electron transfer (DIET) therefore may primarily occur, either by direct contact between cells (*1*), or through nanowires (*2*), or EPS or minerals (*3*), or shuttled by soluble mediators (*4*)

3 Anaerobic Granular Sludge Ecology

The efficiency of granular sludge for high-rate anaerobic wastewater treatment is established since a long time, but the process of anaerobic granulation is still not well understood despite extensive research into granulation mechanisms, microbial ecology, and other significant contributing factors. [25]. Theories of granulation include amongst other, the interactions and spatial requirements of obligate syntrophic groups. One of the limitations in the study of granules is the relatively high abundance of unknown and uncultivated strains, that restrains genomic or proteomic studies of this particular biofilm. The cells proximity that facilitates the electron transfer undoubtedly plays an important role [26]. Most often, when the primary substrate hydrolysis is not the limiting step, granules are built as a layered structure (Fig. 1) that conforms to the sequential anaerobic degradation of organic matter: hydrolysis and acidogenesis, acetogenesis, and methanogenesis [15, 27, 28]. Fluorescent in situ hybridization (FISH) analysis shows that members of *Chloroflexi* and *Betaproteobacteria* colonized mainly the outer layer, members of *Firmicutes* occupied mainly the middle layer [29] and *Methanosaeta*-, *Methanobacterium*-, *Methanospirillum*-, and *Methanosarcina*-like cells are prevalent in the innermost layer [30, 31]. Aging at as short as 4 weeks [31], or being substrate limited [29], the center of the granule tends to be inactive and composed of inert matter, either dormant or decaying microbial cells, and inorganic materials [30]. Even though this layered architecture is coarse, confocal laser scanning microscopy (CLSM) and FISH pictures clearly show large areas in the middle layer that are simultaneously colonized by the bacterial and archaeal domains, that likely correspond syntrophic populations [30, 32].

Some studies demonstrate evolution and changing characteristics of the granule population over time [31]. There are likely species, and even sub-species and mutants that are advantaged by the selective pressure exerted in this kind of granule environment. One avenue to understand the evolution of species in the anaerobic granule is the use of simplified models using only a few pure strains. Summers and coworkers [33] show that in a syntrophic coculture of *Geobacter metallireducens* and *G. sulfurreducens* that forms aggregates similar to granules, a single mutation in PilR gene always appears in *G. sulfurreducens* (PilR is an enhancer binding protein that plays a role in transcriptional regulation when coupled with RpoN, an RNA polymerase). This mutation inactivates the binding function of PilR and up- or down-regulates the expression of several genes. Globally, this mutation in PilR seems to facilitate aggregation and syntrophic metabolism. Interestingly, OmcS, a c-type cytochrome, was over-expressed in PilR mutants [34] and was abundantly found extracellularly in aggregated cocultures of *G. metallireducens* and *G. sulfurreducens*, as revealed by immunogold labelling [33]. OmcS is a central protein involved in the reduction of extracellular, insoluble Fe(III) oxides. It should be noted that all others cytochromes, also involved in extracellular transport of electrons were down-regulated in PilR mutants [34] and in *G. sulfurreducens* cells growing with an anode as electron acceptor [35]. Since there is no need to over-express and produce cytochromes since the pathway of iron reduction is not involved in this assay, why would cytochromes be so important? Cytochromes are known to be involved in the electron transfer and the fact they are abundantly present outside the cell suggests that DIET may occur. Granules have much more complex populations than only *Geobacteraceae* species, however, as in all other ecosystems, the evolutionary pressure in a granule will select for most adapted cells. Individual cells that were disadvantaged, by dispersion in a liquid suspension could have a competitive advantage in the granule cluster, in particular if they could better compete for energy due to DIET.

Another interesting case that increases our understanding is the coculture ecology of *Pelotomaculum thermopropionicum* and *Methanothermobacter thermautotrophicus*. Together these two species form aggregates that facilitate the interspecies electron transfer and increase the growth rate. What is particularly interesting is that the flagellum-like filaments of *P. thermopropionicum* connect to *M. thermautotrophicus* cells to finally wrap them in aggregates [14]. It was then shown that these filaments were electrically conductive [17]. Despite these results, there is still no direct experimental evidence that the fermentative bacterium *P. thermopropionicum* directly transfers electrons to *M. thermautotrophicus* with those filaments. Recently, Shimoyama and coworkers found a protein-mediated interaction between filaments and the *M. thermautotrophicus* surface [36]. The flagellar cap protein FliD, that is part of the filament structure of *P. thermopropionicum*, was found to bind to the syntrophic partner cell surface, *M. thermautotrophicus* and *Methanosaeta thermophila*. Binding of FliD to the surface of *M. thermautotrophicus* induces a change in the expression of over 50 genes and most of them have a positive impact on the syntrophic metabolism. This is another example of the specific association between syntrophic microorganisms, and we can expect that several other interactions of this type could occur in methanogenic environments. The hypothesis that flagellum-like proteins binding to methanogenic cells allow them to capture directly electrons from DIET is supported since these filaments are known to be conductive. Granulation seems to give an advantage to organisms, allowing for the development of efficient strategies of interspecies electron transfer and this pressure could push towards the development of DIET.

Horizontal gene transfer also influences the evolution of granules. By comparing the genome of syntrophic bacteria with gene banks related to methanogenesis, Kato and Watanabe [37] found that the gene content of syntrophic bacteria is more similar to methanogens with which they interact than with other phylogenetically close microorganisms. This genomic similitude with the microbial neighbourhood also includes genes from their central metabolism such as hydrogenases or ATPases. Codon-usage patterns of genes were also more influenced by the microorganisms at proximity than the phylogeny. This niche-associated evolution probably has an important impact on granule development. Indeed, environment (including neighbour microbes) could further influence the population genetics and function than previously thought. Adaptive genes coding for DIET in granules could then efficiently spread to several genetically distant species.

4 Exoelectrogenic Bacteria: The First Partner

Enrichment of electricity-producing microorganisms (or exoelectrogens) could be obtained from placing a graphite anode in marine sediments [38] and most of the time it is mainly composed by members of *Geobacteraceae* family [39]. However, many other exoelectrogenic bacteria have been found widespread in the bacteria phylogenetic tree, even though most pertain to the *Proteobacteria* phylum [19]. Exoelectrogens are ubiquitous in anaerobic environments, natural in sediment, or engineered for anaerobic digesters. As a corroboration, anaerobic sludge is often used as inoculum in BES. It seems that exoelectrogens first developed pathways to achieve the dissimilatory reduction of extracellular compounds such as metal oxides. In *G. sulfurreducens*, a membrane NADH-dehydrogenase initiates the exportation of electron across the membrane through a chain of cytochromes to reach the outer membrane layer [16]. *Shewanella oneidensis* use a similar pathway to export electron with a membrane formate-dehydrogenase [40]. It was also found that some strains could secrete cytochromes or other electron-shuttling compounds in the environment to more easily reach distant solid electron acceptors [41, 42]. Some strains such as *P. thermopropionicum*, *S. oneidensis* and the cyanobacterium

Synechocystis have developed nanowires (flagellum-like filaments) to reach further electron acceptors and those nanowires seem to be specialized in electron transport over longer distances, such as 20 μm [17, 43].

Interspecies electron transfer during anaerobic digestion is already well characterized [6, 44]. The microbial species that provide electrons in this exchange are the hydrogen and formate producers. Therefore, those same species should be the ones that supply electrons in DIET also. That includes almost all bacteria, which perform dark fermentation. Specifically, in this large group, the candidates for DIET should be those who are genetically equipped to perform this kind of exchange because of their ability to carry out extra-cellular electron export. However, in granules, syntrophic bacteria are the most likely to undergo a selective pressure inducing the development for DIET capability, because of the thermodynamic limitations related to the hydrogen partial pressure (or formate concentration) and the absolute need to have an electron sink. As a result, syntrophic bacteria represent the best candidates for performing exoelectrogenesis and DIET.

Alike with the traditional interspecies electron transfer mediated by hydrogen or formate, DIET should be thermodynamically advantageous for both electron donor and acceptor. The most common reduced molecules formed during metabolism of syntrophic or exoelectrogenic bacteria are NADH and reduced ferredoxin with standard redox potentials ($E°'$) of −320 and −420 mV respectively. Several hydrogenases can oxidize these reduced compounds to produce hydrogen ($E°' = -414$ mV). However, the hydrogen partial pressure must stay very low (<10 Pa), otherwise the reduced compounds accumulate and inhibit growth. Finding an electron sink is thus vital for syntrophic bacteria. On the other hand, the first step of the methanogenesis is to reduce the ferredoxin ($E°' = -500$ mV), which is typically made by the hydrogen oxidation. As this reaction is endergonic, methanogens use energy coming from the chemiosmotic H^+ gradient created across the cell membrane with further exergonic methanogenic reactions catalyzed by energy-converting hydrogenases [45, 46]. The final step of methanogenesis leads to methane, which has an $E°'$ of −240 mV.

5 Electrotrophic Methanogens: The Second Partner

To achieve DIET, exoelectrogens need partners that are able to accept electrons to generate their energy, such as electrotrophic microorganisms. In 1987, Daniels and coworkers observed that methanogens cultivated with elemental iron (Fe^o) as the only source of electrons were able to reduce CO_2 to produce methane [47, 48]. This means electrons were imported through the cell membrane. Afterwards, it was found that some *Methanobacterium*-like and *Methanococcus* archaea were able to produce methane faster with iron than other well-known hydrogen-consuming methanogens [49, 50]. It was also shown that the supply of solid iron particles increased the methane production in anaerobic sludge compared to endogenous substrate controls [51]. This suggests a more direct pathway to an electron source than hydrogen or formate, and this pathway could represent a competitive advantage for energy in the anaerobic granular sludge. In the same vein, it was shown that a cathode can serve as a direct electron donor to methanogens [52].

Electrotrophs have two ways to produce energy with directly transferred electrons: simply produce hydrogen, or integrate them directly in the electron transport chain to produce other reduced compounds and chemiosmotic energy. In the first case, hydrogenases would generate hydrogen that would be directly used by the methanogen. In the second case, some membrane-associated hydrogenases (for example EchA-F, VhoACG, HdrDE, etc.) in methanogens would directly reduce ferredoxin, methanophenazine or CoM-S-S-CoB, using free electrons [53–56]. At the same time, those enzymatic complexes could generate a proton or sodium gradient across the membrane, a useful source of energy for methanogens. Reduced compounds will be finally used to form methane, or assimilated in biosynthesis, or even yet used to form hydrogen, that can be used by some hydrogenases in methanogens. Only *Methanosarcinales* have cytochromes [54]. But if these membrane-bound hydrogenases play a role in the transport of electrons across membranes [53], they might then accept electrons directly and methanogens other than *Methanosarcinales* might perform DIET as well.

During evolution, *Methanosarcinales* developed the ability to produce methane from CO_2 reduction as well as from acetate dismutation. The only exception is the genus *Methanosaeta*, whose members were known to be unable to reduce CO_2. Recently, DIET was demonstrated with *Methanosaeta harundinacea* [32]. *M. harundinacea* was grown in coculture with *G. metallireducens*, a proteobacterium able to oxidize ethanol with Fe^{3+} as the only electron acceptor, but unable to produce hydrogen or formate. However, *M. harundinacea*, in association with the ethanol-consuming *G. metallireducens*, was able to produce methane with an electron recovery efficiency of 96 ± 2 %, meaning that not only acetate resulting from the ethanol oxidation was used, but also residual electrons and exogenous CO_2 [32]. Those authors found that genes related to CO_2 reduction to methane were highly expressed in such a coculture.

Methanosarcina barkeri was also recently found to have DIET capability when associated with *G. metallireducens* [57, 58]. Interestingly, the direct use of electrons by *M. barkeri* does not inhibit the acetoclastic pathway, which was the case when hydrogen was present [59]. The ability of processing acetate at the same time as reducing CO_2 represents a competitive advantage for those methanogens that can be involved in DIET.

Recently, one strain of *Sporomusa* sp. was shown to grow as an acetogen with Fe(0) as the sole electron donor and to enhance iron corrosion, which was the first demonstration that bio-corrosion can be mediated by pure acetogenic bacteria [60]. This strain is thus able to take electrons directly from an extracellular donor to produce acetate: similar mechanism is also plausible in anaerobic granules.

6 A New Role for the Granule Extracellular Polymeric Substances (EPS): Electron Dispatcher

In granule, mixed populations are surrounded by dead cells, debris and inert matter, embedded in extracellular polymeric substances (EPS), that constitute the matrix of biofilms [61]. EPS contain proteins, polysaccharides, enzymes, humic substances, lipids, nucleic acids, other complex organic molecules and minerals and can represent up to 20 % of the dry weight of a biofilm. The EPS take form and change over time. As EPS are mostly synthesized by nearby cells, their characteristics and properties are highly related to the populations they harbour. In a column of sand (approximately 10 cm) that was colonized by *S. oneidensis* MR-1, Ntarlagiannis and coworkers [62] showed that the bacteria synthesized nanowires all around each sand grain "hardwiring" the column bottom up, so that oxygen, the terminal electron acceptor, could be reached at the top. Even when there is a distance in cm between the electron donors and acceptors, cells can develop mechanisms to overcome the distance obstacle. This example shows how cells can impact their surrounding EPS, create and modify them to their advantage.

The interspecies electron exchange depends on direct or indirect contacts between cells, which must be stable and durable. With the direct type of contact the exoelectrogen's membrane touches the electrotroph's membrane and electrons have just to cross the two membranes. That is the optimal way to perform DIET. But in lab-scale experiments, DIET has only been observed with cocultures, which develop as aggregates alike anaerobic granules, and it is likely that contact between cells can only be indirect. Indirect contact implies vectors to conduct electrons from one to another cell and those vectors should be part of the EPS. They were first evidenced in studies on dissimilatory metal oxides [63]. Then, they were given a role of electron shuttles with the development of BES systems. Electron shuttles could be humic substances, riboflavins, cystein, sulphide, phenazine, anthrahydroquinone disulfonate (AQDS), etc. [11]. Those molecules could be reduced and oxidized with relative ease, compared to metal oxides, facilitating the electron exchange from a donor cell to a receptor one (or an electrode). Cell components or molecules such as flagella and nanowires, cytochromes, pilin, flavins and even DNA are all candidates to play a role in direct electron transfer in biofilms [17, 34, 36, 43, 64–66].

Another vector possibility for electron transfer is to use other cells, alive or dead. Nielsen and coworkers showed in sea sediments, that hydrogen sulphide and organic matter that are sufficiently deep so to not have access to oxygen at the surface, were nevertheless oxidized because electrons could travel up to 12 mm across the sediment upper layer [67, 68]. They also demonstrated that the native conductors were long, filamentous bacteria of the *Desulfobulbaceae* family, able to grow and elongate fast in vertically expanded filaments, and to transport electrons across centimeter-long distances [69]. Conductive minerals also seem to play a role in the sediment conductivity [70]. On the other hand, Esteve-Nunez and coworkers [71] showed that *G. sulfurreducens* had externally-bound cytochromes so that the cell could act as electron-accepting capacitors and could metabolically function during 8 min without electron acceptors in the medium. In total, 10^7 electrons could be accepted per cell. Within an EPS-dense biofilm, it is realistic to imagine that electrons transit from an exoelectrogenic cell to other cell walls to finally reach the membrane of an electrotrophic cell. Likewise, debris of dead cells could also be used to transfer electrons to higher redox zones.

Anaerobic granules have been found electrically conductive. This conductance does not seem to be related to minerals such as iron contained in the granules, since the temperature dependence response of the aggregate conductance was characteristic of the organic metallic-like conductance previously described for the pili of *G. sulfurreducens* and was inconsistent with minerals conductance [72]. Incidentally a moderate correlation was found ($r = 0.67$) between the abundance of *Geobacter* species in the UASB granules and granule conductivity, suggesting that *Geobacter* contributed to granule conductivity [73]. In contrast, mineral particles such as magnetite, carbon cloth, granular activated carbon, and biochar have been shown able to facilitate electron transfer, presumably DIET [58, 74–76]. Conductive particles were even shown experimentally that they could replace EPS as biological vectors after anaerobic granules were disintegrated and washed [77]. Although the use of minerals to increase the anaerobic reactors efficiency was known for a long time, it is only recently that DIET is seen as one mechanism behind these positive impacts in anaerobic digestion.

Finally, electron vectors should have a redox potential (E′) so that DIET is thermodynamically feasible, i.e. a potential that results in a negative change in Gibbs free energy ($\Delta G'$). Vectors in EPS must hence have a higher E′ than that of the reduced compounds formed during the substrate oxidation and a lower E′ than the ultimate electron acceptor. As an example, the $\Delta G'$ of NADH oxidation coupled to an electron vector reduction is given by:

$$\Delta G' = \Delta G^{\circ\prime} + R \cdot T \cdot \ln \frac{[\mathrm{NAD}^+] \cdot [\mathrm{vector}^-]}{[\mathrm{NADH}] \cdot [\mathrm{vector}^+]}$$

For $\Delta G'$ to be negative, as in the case of the molecular hydrogen (H_2) transfer, the concentration of reduced vectors should be low. If electrotrophic methanogens capture electrons as fast as the vectors are reduced, cells would be able to regenerate their NADH as they would be permanently surrounded by oxidized vector molecules.

7 Conclusion

DIET is a recently discovered form of extracellular electron transfer in which microorganisms exchange electrons to cooperatively degrade organic compounds under anaerobic conditions. The phenomenon might also happen in anoxic or aerobic environments, with terminal electron acceptors other than CO_2, such as sulfate, nitrate, protons or oxygen. This would permit, for example, aerobic cells that live deeply in biofilms to metabolize their substrate even though they have difficulty in accessing oxygen, since O_2 diffusion is limited. DIET could be a more generalized way of getting energy for microbes than it was previously thought, being effected in a large variety of environments.

DIET is potentially an important mechanism for electron exchange in the anaerobic consortia, involved in the conversion of organic wastes into methane, particularly anaerobic granules used in the wastewater treatment (Fig. 1). Though, there has not yet been any direct experimental observation of DIET within such granules, and indirect clues of DIET relate only to *Geobacter*-enriched granules from brewery anaerobic wastewater treatment plants. Hence, there is still much work needed to understand this phenomenon. Genomic and proteomic studies on the complex populations of multispecies biofilms and chemical and physical characterizations of EPS should advance the DIET evidence and understanding in the near future.

A better mechanistic understanding of DIET will allow for technology improvement in the anaerobic digestion field, such as based on the artificial enhancement of DIET.

References

1. Potter MC (1911) Electrical effects accompanying the decomposition of organic compounds. Proc R Soc Lond B 84:260–276
2. Arends JBA, Verstraete W (2011) 100 years of microbial electricity production: three concepts for the future. Microb Biotechnol 5:333–346
3. Logan BE, Regan JM (2006) Electricity-producing bacterial communities in microbial fuel cells. Trends Microbiol 14:512–518
4. Lovley DR (2006) Microbial fuel cells: novel microbial physiologies and engineering approaches. Curr Opin Biotechnol 17:327–332
5. Kurakin A (2011) The self-organizing fractal theory as a universal discovery method: the phenomenon of life. Theor Biol Med Model 8:1–66
6. Stams AJM, Plugge CM (2009) Electron transfer in syntrophic communities of anaerobic bacteria and archaea. Nat Rev Microbiol 9:568–577
7. Bryant MP, Wolin EA, Wolin MJ, Wolfe RS (1967) *Methanobacillus omelianskii*, a symbiotic association of two species of bacteria. Arch Mikrobiol 59:20–31
8. Thiele JH, Zeikus JG (1988) Control of interspecies electron flow during anaerobic digestion: significance of formate transfer versus hydrogen transfer during syntrophic methanogenesis in flocs. Appl Environ Microbiol 54:20–29
9. Boone DR, Johnson RL, Liu Y (1989) Diffusion of the interspecies electron carriers H_2 and formate in methanogenic ecosystems and its implications in the measurement of K_m for H_2 or formate uptake. Appl Environ Microbiol 55:1735–1741
10. Schink B (2008) Energetic aspects of methanogenic feeding webs. In: Wall JD, Harwood CS, Demain AL (eds) Bioenergy. ASM Press, Washington, pp 171–178
11. Stams AJM, De Bok FAM, Plugge CM, Van Eekert MHA, Dolfing J, Schraa G (2006) Exocellular electron transfer in anaerobic microbial communities. Environ Microbiol 8: 371–382

12. Lettinga G (1995) Anaerobic digestion and wastewater treatment systems. Antonie Van Leeuwenhoek 67:3–28
13. Lettinga G, van Velsen AFM, Hobma SW, de Zeeuw W, Klapwijk A (1980) Use of the upflow sludge blanket (USB) reactor concept for biological wastewater treatment, especially for anaerobic treatment. Biotechnol Bioeng 22:699–734
14. Ishii S, Kosaka T, Hori K, Hotta Y, Watanabe K (2005) Coaggregation facilitates interspecies hydrogen transfer between *Pelotomaculum thermopropionicum* and *Methanothermobacter thermautotrophicus*. Appl Environ Microbiol 71:7838–7845
15. Guiot SR, Pauss A, Costerton JW (1992) A structured model of the anaerobic granule consortium. Water Sci Technol 25(7):1–10
16. Lovley DR, Holmes DE, Nevin KP (2004) Dissimilatory Fe(III) and Mn(IV) reduction. Adv Microb Physiol 49:219–286
17. Gorby YA, Yanina S, McLean JS, Rosso KM, Moyles D, Dohnalkova A, Beveridge TJ, Chang IS, Kim BH, Kim KS, Culley DE, Reed SB, Romine MF, Saffarini DA, Hill EA, Shi L, Elias DA, Kennedy DW, Pinchuk G, Watanabe K, Ishii S, Logan BE, Nealson KH, Fredrickson JK (2006) Electrically conductive bacterial nanowires produced by *Shewanella oneidensis* strain MR-1 and other microorganisms. Proc Natl Acad Sci U S A 103:11358–11363
18. Cheng S, Xing D, Call DF, Logan BE (2009) Direct biological conversion of electrical current into methane by electromethanogenesis. Environ Sci Technol 43:3953–3958
19. Logan BE (2009) Exoelectrogenic bacteria that power microbial fuel cells. Nat Rev Microbiol 7:375–381
20. Nealson KH (2010) Geomicrobiology: Sediment reactions defy dogma. Nature 463: 1033–1034
21. Lovley DR, Nevin KP (2011) A shift in the current: New applications and concepts for microbe-electrode electron exchange. Curr Opin Biotechnol 22:441–448
22. Pauss A, Samson R, Guiot SR (1990) Thermodynamic evidence of trophic microniches in methanogenic granular sludge-bed reactors. Appl Microbiol Biotechnol 33:88–92
23. Guiot SR, MacLeod FA, Pauss A (1990) Thermodynamical and microbiological evidence of trophic microniches for propionate degradation in a methanogenic sludge-bed reactor. In: Belaich J-P, Bruschi M, Garcia J-L (eds) Microbiology and biochemistry of strict anaerobes involved in interspecies hydrogen transfer. Federation of European Microbiological Societies Symposium Series, vol 54. Plenum Press, New York, pp 173–183
24. Frigon J-C, Guiot SR (1995) Impact of liquid-to-gas hydrogen mass transfer on substrate conversion efficiency of a upflow anaerobic sludge bed and filter reactor. Enzyme Microbiol Technol 17:1080–1086
25. Picioreanu C, Batstone DJ, van Loosdrecht MCM (2005) Multidimensional modelling of anaerobic granules. Water Sci Technol 52:501–507
26. Ishii S, Kosaka T, Hotta Y, Watanabe K (2006) Simulating the contribution of coaggregation to interspecies hydrogen fluxes in syntrophic methanogenic consortia. Appl Environ Microbiol 72:5093–5096
27. MacLeod FA, Guiot SR, Costerton JW (1990) Layered structure of bacterial aggregates produced in an upflow anaerobic sludge bed and filter reactor. Appl Environ Microbiol 56:1298–1307
28. Rocheleau S, Greer CW, Lawrence JR, Cantin C, Laramée L, Guiot SR (1999) Differentiation of *Methanosaeta concilii* and *Methanosarcina barkeri* in anaerobic mesophilic granular sludge by using fluorescent in situ hybridization and confocal scanning laser microscopy. Appl Environ Microbiol 65:2222–2229
29. Satoh H, Miura Y, Tsushima I, Okabe S (2007) Layered structure of bacterial and archaeal communities and their in situ activities in anaerobic granules. Appl Environ Microbiol 73:7300–7307
30. Sekiguchi Y, Kamagata Y, Nakamura K, Ohashi A, Harada H (1999) Fluorescence in situ hybridization using 16S rRNA-targeted oligonucleotides reveals localization of methanogens and selected uncultured bacteria in mesophilic and thermophilic sludge granules. Appl Environ Microbiol 65:1280–1288

31. Diaz EE, Stams AJM, Amils R, Sanz JL (2006) Phenotypic properties and microbial diversity of methanogenic granules from a full-scale upflow anaerobic sludge bed reactor treating brewery wastewater. Appl Environ Microbiol 72:4942–4949
32. Rotaru A-E, Shrestha PM, Liu F, Shrestha M, Shrestha D, Embree M, Zengler K, Wardman C, Nevin KP, Lovley DR (2014) A new model for electron flow during anaerobic digestion: direct interspecies electron transfer to *Methanosaeta* for the reduction of carbon dioxide to methane. Energy Environ Sci 7:408–415
33. Summers ZM, Fogarty HE, Leang C, Franks AE, Malvankar NS, Lovley DR (2010) Direct exchange of electrons within aggregates of an evolved syntrophic coculture of anaerobic bacteria. Science 330:1413–1415
34. Juárez K, Kim BC, Nevin K, Olvera L, Reguera G, Lovley DR, Methé BA (2009) PilR, a transcriptional regulator for pilin and other genes required for Fe(III) reduction in *Geobacter sulfurreducens*. J Mol Microbiol Biotechnol 16:146–158
35. Holmes DE, Chaudhuri SK, Nevin KP, Mehta T, Methé BA, Liu A, Ward JE, Woodard TL, Webster J, Lovley DR (2006) Microarray and genetic analysis of electron transfer to electrodes in *Geobacter sulfurreducens*. Environ Microbiol 8:1805–1815
36. Shimoyama T, Kato S, Ishii S, Watanabe K (2009) Flagellum mediates symbiosis. Science 323:1574
37. Kato S, Watanabe K (2010) Ecological and evolutionary interactions in syntrophic methanogenic consortia. Microbes Environ 25:145–151
38. Reimers CE, Tender LM, Fertig S, Wang W (2001) Harvesting energy from the marine sediment-water interface. Environ Sci Technol 35:192–195
39. Bond DR, Holmes DE, Tender LM, Lovley DR (2002) Electrode-reducing microorganisms that harvest energy from marine sediments. Science 295:483–485
40. Tikhonova TV, Popov VO (2014) Structural and functional studies of multiheme cytochromes c involved in extracellular electron transport in bacterial dissimilatory metal reduction. Biochemistry (Moscow) 79:1584–1601
41. Nevin KP, Lovley DR (2002) Mechanisms for accessing insoluble Fe(III) oxide during dissimilatory Fe(III) reduction by *Geothrix fermentans*. Appl Environ Microbiol 68:2294–2299
42. Seeliger S, Cord-Ruwisch R, Schink B (1998) A periplasmic and extracellular c-type cytochrome of Geobacter sulfurreducens acts as a ferric iron reductase and as an electron carrier to other acceptors or to partner bacteria. J Bacteriol 180:3686–3691
43. Reguera G, McCarthy KD, Mehta T, Nicoll JS, Tuominen MT, Lovley DR (2005) Extracellular electron transfer via microbial nanowires. Nature 435:1098–1101
44. Schmidt JE, Ahring BK (1996) Granular sludge formation in upflow anaerobic sludge blanket (UASB) reactors. Biotechnol Bioeng 49:229–246
45. Thauer RK, Kaster AK, Goenrich M, Schick M, Hiromoto T, Shima S (2010) Hydrogenases from methanogenic archaea, nickel, a novel cofactor, and H2 storage. Ann Rev Biochem 79:507–536
46. Kaster A-K, Goenrich M, Seedorf H, Liesegang H, Wollherr A, Gottschalk G, Thauer RK (2011) More than 200 genes required for methane formation from H_2 and CO_2 and energy conservation are present in *Methanothermobacter marburgensis* and *Methanothermobacter thermautotrophicus*. Archaea 11:1–23
47. Daniels L, Belay N, Rajagopal BS, Weimer PJ (1987) Bacterial methanogenesis and growth from CO_2 with elemental iron as the sole source of electrons. Science 237:509–511
48. Chastain BK, Kral TA (2010) Zero-valent iron on Mars: an alternative energy source for methanogens. Icarus 208:198–201
49. Dinh HT, Kuever J, Mußmann M, Hassel AW, Stratmann M, Widdel F (2004) Iron corrosion by novel anaerobic microorganisms. Nature 427:829–832
50. Uchiyama T, Ito K, Mori K, Tsurumaru H, Harayama S (2010) Iron-corroding methanogen isolated from a crude-oil storage tank. Appl Environ Microbiol 76:1783–1788
51. Karri S, Sierra-Alvarez R, Field JA (2005) Zero valent iron as an electron-donor for methanogenesis and sulfate reduction in anaerobic sludge. Biotechnol Bioeng 92:810–819

52. Gregory KB, Bond DR, Lovley DR (2004) Graphite electrodes as electron donors for anaerobic respiration. Environ Microbiol 6:596–604
53. Rosenbaum M, Aulenta F, Villano M, Angenent LT (2011) Cathodes as electron donors for microbial metabolism: Which extracellular electron transfer mechanisms are involved? Bioresour Technol 102:324–333
54. Thauer RK, Kaster A-K, Seedorf H, Buckel W, Hedderich R (2008) Methanogenic archaea: ecologically relevant differences in energy conservation. Nat Rev Microbiol 6:579–591
55. Welte C, Kallnik V, Grapp M, Bender G, Ragsdale S, Deppenmeier U (2010) Function of Ech hydrogenase in ferredoxin-dependent, membrane-bound electron transport in Methanosarcina mazei. J Bacteriol 192:674–678
56. Gottschalk G, Blaut M (1990) Generation of proton and sodium motive forces in methanogenic bacteria. BBA-Bioenerg 1018:263–266
57. Rotaru A-E, Shrestha PM, Liu F, Markovaite B, Chen S, Nevin KP, Lovley DR (2014) Direct interspecies electron transfer between *Geobacter metallireducens* and *Methanosarcina barkeri*. Appl Environ Microbiol 80:4599–4605
58. Chen S, Rotaru AE, Liu F, Philips J, Woodard TL, Nevin KP, Lovley DR (2014) Carbon cloth stimulates direct interspecies electron transfer in syntrophic co-cultures. Bioresour Technol 173:82–86
59. Ferguson TJ, Mah RA (1983) Effect of H_2-CO_2 on methanogenesis from acetate or methanol in *Methanosarcina* spp. Appl Environ Microbiol 46:348–355
60. Kato S, Yumoto I, Kamagata Y (2015) Isolation of acetogenic bacteria that induce biocorrosion by utilizing metallic iron as the sole electron donor. Appl Environ Microbiol 81:67–73
61. Baloch MI, Akunna JC, Kierans M, Collier PJ (2008) Structural analysis of anaerobic granules in a phase separated reactor by electron microscopy. Bioresour Technol 99:922–929
62. Ntarlagiannis D, Atekwana EA, Hill EA, Gorby Y (2007) Microbial nanowires: Is the subsurface "hardwired"? Geophys Res Lett 34:1–5
63. Lovley DR (1993) Dissimilatory metal reduction. Annu Rev Microbiol 47:263–290
64. Busalmen JP, Esteve-Núñez A, Berná A, Feliu JM (2008) C-type cytochromes wire electricity-producing bacteria to electrodes. Angew Chem Int Ed Engl 47:4874–4877
65. Cohen H, Nogues C, Naaman R, Porath D (2005) Direct measurement of electrical transport through single DNA molecules of complex sequence. Proc Nat Acad Sci USA 102:11589–11593
66. Marsili E, Baron DB, Shikhare ID, Coursolle D, Gralnick JA, Bond DR (2008) *Shewanella* secretes flavins that mediate extracellular electron transfer. Proc Natl Acad Sci USA 105:3968–3973
67. Nielsen LP, Risgaard-Petersen N, Fossing H, Christensen PB, Sayama M (2010) Electric currents couple spatially separated biogeochemical processes in marine sediment. Nature 463:1071–1074
68. Schauer R, Risgaard-Petersen N, Kjeldsen KU, Tataru Bjerg JJ, Jørgensen BB, Schramm A, Nielsen LP (2014) Succession of cable bacteria and electric currents in marine sediment. ISME J 8:1314–1322
69. Pfeffer C, Larsen S, Song J, Dong M, Besenbacher F, Meyer RL, Kjeldsen KU, Schreiber L, Gorby YA, El-Naggar MY, Leung KM, Schramm A, Risgaard-Petersen N, Nielsen LP (2012) Filamentous bacteria transport electrons over centimetre distances. Nature 491:218–221
70. Malvankar NS, King GM, Lovley DR (2015) Centimeter-long electron transport in marine sediments via conductive minerals. ISME J 9:527–531
71. Esteve-Núñez A, Sosnik J, Visconti P, Lovley DR (2008) Fluorescent properties of c-type cytochromes reveal their potential role as an extracytoplasmic electron sink in *Geobacter sulfurreducens*. Environ Microbiol 10:497–505
72. Morita M, Malvankar NS, Franks AE, Summers ZM, Giloteaux L, Rotaru AE, Rotaru C, Lovley DR (2011) Potential for direct interspecies electron transfer in methanogenic wastewater digester aggregates. mBio 2:e00159–11

73. Shrestha PM, Malvankar NS, Werner JJ, Franks AE, Rotaru A-E, Shrestha M, Liu F, Nevin KP, Angenent LT, Lovley DR (2014) Correlation between microbial community and granule conductivity in anaerobic bioreactors for brewery wastewater treatment. Bioresour Technol 174:306–310
74. Cruz Viggi C, Rossetti S, Fazi S, Paiano P, Majone M, Aulenta F (2014) Magnetite particles triggering a faster and more robust syntrophic pathway of methanogenic propionate degradation. Environ Sci Technol 48:7536–7543
75. Liu F, Rotaru AE, Shrestha PM, Malvankar NS, Nevin KP, Lovley DR (2012) Promoting direct interspecies electron transfer with activated carbon. Energy Environ Sci 5:8982–8989
76. Chen S, Rotaru AE, Shrestha PM, Malvankar NS, Liu F, Fan W, Nevin KP, Lovley DR (2014) Promoting interspecies electron transfer with biochar. Sci Rep 4:5019
77. Dubé C-D, Guiot SR (2014) Direct interspecies electron transfer investigation in granular sludge. In: Gübitz GM, Gronauer A, Bauer A, Bochman G, Scheidl S, Weiß S (eds) Int Conf on Anaerobic Digestion, Biogas Science 2014, Vienna, Austria, 2014. University of Natural resources and Life Sciences, Vienna, Austria, pp 85–86

A Critical Assessment of Microbiological Biogas to Biomethane Upgrading Systems

Simon K.-M.R. Rittmann

Abstract Microbiological biogas upgrading could become a promising technology for production of methane (CH_4). This is, storage of irregular generated electricity results in a need to store electricity generated at peak times for use at non-peak times, which could be achieved in an intermediate step by electrolysis of water to molecular hydrogen (H_2). Microbiological biogas upgrading can be performed by contacting carbon dioxide (CO_2), H_2 and hydrogenotrophic methanogenic Archaea either in situ in an anaerobic digester, or ex situ in a separate bioreactor. In situ microbiological biogas upgrading is indicated to require thorough bioprocess development, because only low volumetric CH_4 production rates and low CH_4 fermentation offgas content have been achieved. Higher volumetric production rates are shown for the ex situ microbiological biogas upgrading compared to in situ microbiological biogas upgrading. However, the ex situ microbiological biogas upgrading currently suffers from H_2 gas liquid mass transfer limitation, which results in low volumetric CH_4 productivity compared to pure H_2/CO_2 conversion to CH_4. If waste gas utilization from biological and industrial sources can be shown without reduction in volumetric CH_4 productivity, as well as if the aim of a single stage conversion to a CH_4 fermentation offgas content exceeding 95 vol% can be demonstrated, ex situ microbiological biogas upgrading with pure or enrichment cultures of methanogens could become a promising future technology for almost CO_2-neutral biomethane production.

Keywords Hydrogen · Carbon dioxide · CH_4 · Bioreactor · Anaerobic digestion · Bioprocess development · Pure culture · Enrichment culture · Biofuel · Methanation

S.K.-M.R. Rittmann (✉)
Archaea Biology and Ecogenomics Division, Department of Ecogenomics and Systems Biology, University of Vienna, Althanstraße 14, 1090 Vienna, Austria
e-mail: simon.rittmann@univie.ac.at

G.M. Guebitz et al. (eds.), *Biogas Science and Technology*,
Advances in Biochemical Engineering/Biotechnology 151,
DOI 10.1007/978-3-319-21993-6_5

Contents

1 Introduction

Renewable energy production and technology development become increasingly important regarding the inevitable forthcoming ending of coal, oil and gas reserves as well as the accelerated accumulation of greenhouse gasses in the Earth's atmosphere. The accumulation of greenhouse gasses in the atmosphere will presumably be tremendously increased in rate through exploitation and burning of non-renewable fossil fuels (natural gas, coal and oil) [1–3]. Exploitation and utilization of fossil fuels will contribute massively to greenhouse gas production upon combustion of fossil fuels to carbon dioxide (CO_2). Another drawback of natural gas (and oil) exploitation is that production sites are located in remote and sometimes disputed areas, which could lead to supply insecurities and dependence on bigger players in the natural gas supplier and exploration/exploitation business. Hithereto, the only demonstrated environmentally friendly renewable gas production technology at industrial scale is anaerobic digestion of biomass originating from agricultural residues and organic waste for biogas (consisting mainly of methane (CH_4) and CO_2)) production [4, 5]. Anaerobic digestion is usually applied at decentralized production sites, and should be considered in a scenario for independent renewable energy production, being equally able to partially replace natural gas production and utilization—if being properly conducted, operated and performed. Moreover, anaerobic digestion is considered being a CO_2-neutral technology, because all the CO_2 emitted through combustion and production of biogas had initially been fixed in biomass.

One drawback of biogas production through anaerobic digestion of biomass is the dilution of CH_4 with CO_2, which results in a low calorimetric value of biogas [6, 7]. Therefore, in order to produce an equal replacement for natural gas, biogas could be microbiologically upgraded to biomethane, whereof the major component is CH_4.

Another challenge to the anaerobic digestion technology for biogas production is the fact that currently mostly decentralized production and consumption of biogas is being performed. Therefore, application of microbiological biogas upgrading, by conversion of residual CO_2 to CH_4, would enable the introduction of almost pure CH_4 into the natural gas grid. This is, biomethane could be inducted into the natural gas grid and transported from decentralized production sites to the consumer through the existing natural gas pipeline and distribution infrastructure. Hence, microbiologically produced biomethane, from remote decentralized production sites, would then be available to the consumers. Furthermore, biogas producers wouldn't have the need for onsite processing of biogas for e.g. electricity generation.

Concerning renewable electricity production (wind and solar electricity) the distribution of electricity during peak production times is currently limited regarding a shortage of power transmission lines. In this regard, renewable energy production processes capable of converting and storing electricity in the form of chemical energy are an interesting prospect, because irregular generated electricity results in a need to store electricity generated at peak times for use at non-peak times. Hence, microbiological biogas upgrading could be one option to store renewably generated electricity via molecular hydrogen (H_2) production (by water electrolysis) and concomitant (bio) methanation of the CO_2 part of biogas by reduction with H_2 to biomethane—a technological process chain often referred to as power-to-gas.

In general, the upgrading of biogas to biomethane can be accomplished by physicochemical separation and disposing of all gaseous compounds other than CH_4 by different methods such as gas/gas or liquid/gas separation by membrane separation, vacuum or pressure swing adsorption, as well as by using cryogenic separation as comprehensively reviewed elsewhere [6, 8].

Typically, biogas is composed of CO_2, hydrogen sulphide (H_2S), water (H_2O), ammonia and other trace compounds, which vary in quantity depending on the substrate used for biogas production. After separation and disposing of CO_2 from CH_4, purified CH_4 could be fed into the natural gas grid. The CO_2, which has been separated from biogas can be upgraded by the reduction with H_2 using a classical chemical reaction referred to as Sabatier process (please refer to literature cited in [9]), which is already carried out at industrial scale. The principle stoichiometric reaction equation for the Sabatier process is:

$$CO_2 + 4H_2 \rightarrow CH_4 + 2H_2O \qquad \Delta G^{\circ} = -130.7\ \text{kJ mol}^{-1}.$$

The Sabatier process has a requirement for pure reactant gasses, high reaction temperatures and putative expensive and rare elements for chemical catalyst production. Due to the sensitivity of the chemical catalyst to extended exposure towards reactant gas impurities (i.e. O_2 and other trace compounds), thorough upstream gas cleaning is required before a reduction of CO_2 with H_2 can be performed. Nevertheless, the chemical catalysts applied in the Sabatier-process degenerate over time, because after biogas purification gaseous trace compounds are still present in the reactant gas, which are detrimental to the chemical catalyst and demand regular replacement [10, 11].

Instead of physicochemical separation of biogas into CH_4 and its by-product compounds, the CO_2 part of biogas can be microbiologically upgraded by reduction of CH_4 with H_2. Microbiological upgrading of CO_2 to biomethane can be performed by each of the three biomethanation technologies, but only in two principal different set-ups (Fig. 1):

(1) In situ microbiological biogas upgrading by adding H_2 into the anaerobic digester
(2) Ex situ microbiological biogas upgrading in a separate bioreactor by contacting H_2, CO_2 and an enrichment culture mainly composed of hydrogenotrophic methanogens
(3) Ex situ microbiological biogas upgrading in a separate bioreactor by contacting H_2, CO_2 and a pure culture of hydrogenotrophic methanogens

From a microbiological point of view upgrading of biogas to biomethane with H_2 is performed by an intriguing group of microorganisms from the domain Archaea, which are referred to as hydrogenotrophic methanogens [12–15]. Hydrogenotrophic methanogenic Archaea are fascinating organisms due to their extraordinary physiological, biochemical and biotechnological features [9, 14–17]. It is well known that hydrogenotrophic methanogens fulfil an important role in the anaerobic digestion process

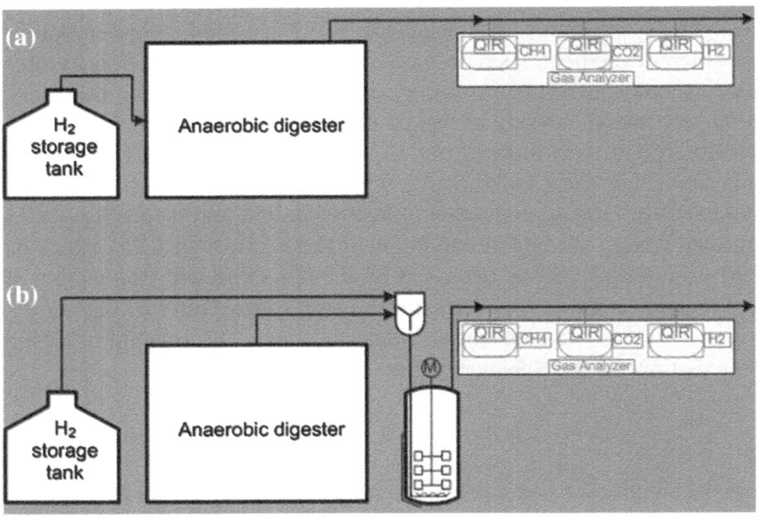

Fig. 1 Two principle schematic set-ups for the microbiological upgrading of biogas to biomethane are presented. H_2 from renewable energy production is converted e.g. *via* water electrolysis (H_2 storage tank). The fermentation offgas needs to be analysed regarding the composition of CH_4, CO_2, H_2 (and putatively also H_2S). **a** shows in situ biogas upgrading by addition of H_2 directly into the anaerobic digester. Due to the simplicity of the set-up a separate bioreactor does not have to be included. In (**b**) the principle set-up for ex situ microbiological biogas upgrading in a separate bioreactor is shown. In this bioreactor H_2, biogas and an enrichment culture comprising mainly of hydrogenotrophic methanogens, or a pure culture of hydrogenotrophic methanogens, can be used for H_2/CO_2 conversion. In this set-up CO_2 from biogas (or also from other CO_2 or H_2 containing industrial flue gasses) can be converted to biomethane under defined process conditions.

[18, 19]. Biomass breakdown and hydrolysis in anaerobic digesters is arranged by a complex consortium of syntrophic growing organisms, including biohydrogen producing and acetogenic microbes, which precede hydrogenotrophic and acetoclastic methanogens in the anaerobic degradation reactions [14]. Besides strict hydrogenotrophic methanogens also facultative hydrogenotrophic methanogens have already been thoroughly characterized [14, 20]. Although methanogens can be discriminated by using biochemical, bioprocess technological, molecular biological and physiological methods [9, 20] the presence or absence of cytochromes can also be used for discriminating different groups of methanogens [14]. The requirement of hydrogenotrophic methanogens containing cytochromes for higher partial pressures of H_2 is at least ten times higher compared to non-cytochrome containing hydrogenotrophic methanogens. This is, methanogens with cytochromes can grow at partial pressures of H_2 as low as 1–10 Pa [14].

Methanation of H_2 and CO_2 for CH_4 production by using pure reactant gasses and microbial monocultures has already received much attention in the past decades, and, as reviewed elsewhere, the cultivation of hydrogenotrophic methanogens has been accomplished using different laboratory-scale bioreactor systems [9]. Quantitative bioprocess development, including scale-up of biomethanation, has only recently become a re-emerging focus [9, 21–23], despite its powerful volumetric CH_4 production rates being known for a long time [24].

Only in a few number of studies the introduction of H_2 into anaerobic digesters at laboratory scale bioreactors was reported for the purpose of in situ microbiological biogas upgrading. Also ex situ microbiological biogas upgrading in a postprocessing step contacting H_2 and biogas in a separate bioreactor containing either a mixed culture of enriched hydrogenotrophic methanogens or a pure culture of hydrogenotrophic methanogens has also only been rarely examined at laboratory scale, as it will be shown in the review sections below.

In this review different lab-scale microbiological biogas upgrading technologies using hydrogenotrophic methanogens are presented. Furthermore, in situ as well as ex situ hydrogenotrophic biomethanation technologies for microbiological biogas upgrading will be quantitatively compared by analysing data available from available publications. First, a recapitulation on the state of the art of the different hydrogenotrophic microbiological biogas upgrading technologies is presented. Thereafter, the pros and cons as well as the industrial readiness of the different technologies will be critically discussed.

2 Review

2.1 In situ Microbiological Biogas Upgrading By Addition of H_2 into Anaerobic Digesters

In situ microbiological biogas upgrading by introducing H_2 into an anaerobic digester has been successfully performed at lab-scale. A 4.5 L continuously stirred tank reactor (CSTR) with 3.5 L working volume containing a thermophilic anaerobic mixed culture was operated at 55 °C [25]. The H_2 gassing rate to the microbiological biogas upgrading bioreactor was 0.0005 volume gas per volume liquid per minute (vvm). H_2 gas flow entered at the bottom of the CSTR through ceramic gas diffusers. The agitation speed was set to 65 rpm. The CSTR was operated for 1.5 months in semi-continuous culture mode. A control bioreactor of the same type was operated under the same conditions but without the addition of H_2. The pH value was not controlled during the cultivation, which resulted in an increase of pH [25], due to stripping of CO_2. As a consequence of higher pH the aceticlastic methanogens might have become negatively affected in volumetric CH_4 productivity. As discussed by the authors, strategies for pH control need to be developed and applied. Furthermore, it seems that H_2 and CO_2 could also have been fixed for acetogenensis [26]. Due to the low H_2 gassing rate, and low agitation speed applied in the bioreactor, only little gas to liquid phase mass transfer of H_2 could be accomplished. Consequently only little CH_4 was produced by hydrogenotrophic methanogens, which is indicated in a low methane evolution rate (MER) of 0.25 mmol L^{-1} h^{-1} (calculated from H_2 consumption rate divided by four). However, combining anaerobic digestion and hydrogenotrophic methanation a total MER of 0.84 mmol CH_4 L^{-1} h^{-1} could be achieved. H_2 gas to liquid phase mass transfer needs to be increased in order to be able to increase MER, because unconverted H_2 and CO_2 still remained in the fermentation offgas, which is reflected in the low CH_4 offgas content of the microbiological biogas upgrading bioreactor comprising 65, 20 and 15 vol% CH_4, H_2 and CO_2, respectively, and compared to 62 and 38 vol% of CH_4 and CO_2 in the control bioreactor, respectively. Clearly a partial exchange of CO_2 to H_2 content in the fermentation offgas has been achieved.

In situ microbiological biogas upgrading of synthetic biogas by addition of H_2 was performed in a 1 L CSTR filled with digested manure to a working volume of 0.6 L. The CSTR was operated at 55 °C at a hydraulic retention time (HRT) of 15 days [27]. A reference bioreactor, operated under identical conditions, except for the addition of H_2, served as control. Different strategies for H_2 gas to liquid mass transfer were applied and the response of two different gas diffusers (column diffuser 0.5–1.0 mm in diameter, or a ceramic diffuser 14–40 mm in diameter), during application of two different agitation speeds (150 and 300 rpm) was investigated. A H_2 gassing rate of 0.0012 vvm was applied. Faster agitation speed was found to increase H_2 solubility and consequently led to better microbiological (synthetic) biogas upgrading. Utilizing the ceramic diffuser led to increased H_2 gas to liquid mass transfer due to smaller bubble size generation, compared to the utilization of the column diffuser for gas addition during fermentation. Highest MER of 0.69 mmol CH_4 L^{-1} h^{-1} (calculated from H_2 consumption rate) was attained by using the column diffuser at an agitation speed of 150 rpm, resulting in a total MER of 1.65 mmol CH_4 L^{-1} h^{-1} obtained by combined CH_4 production from anaerobic digestion as well as from in situ microbiological biogas upgrading. Incomplete conversion of introduced H_2 to CH_4 and partial exchange of the remaining CO_2 with H_2 could not be avoided, which was obviously due to the limited ability of the applied system for quantitative H_2 gas to liquid mass transfer. Complete conversion of CO_2 and H_2 is, however, demanded for an envisioned industrial application of the technology. Nevertheless, long term operation over a period of 150 days under pH controlled conditions could be achieved in semi-continuous culture mode.

2.2 *Ex situ Microbiological Biogas Upgrading in a Separate Bioreactor By Contacting H_2, Biogas and An Enrichment Culture Including Hydrogenotrophic Methanogens*

An enriched hydrogenotrophic and methanogenic mixed culture was established for microbiological upgrading of synthetic biogas, by pre-incubating and adapting a microbial consortium for half a year in closed batch mode with H_2/CO_2 (ratio of 4:1) gassing only [7]. For upgrading experiments a 1 L bottle with 0.6 L working volume was used as bioreactor. The bioreactor was inoculated with the pre-conditioned hydrogenotrophic and methanogenic enrichment culture and the suspension was stirred at either 500 or 800 rpm, respectively. The test gas resembled H_2-enriched biogas of the following composition: 60:25:15 H_2:CH_4:CO_2, respectively. All experiments were performed in semi-continuous culture mode. High offgas quality of 94.2 vol% CH_4 was achieved in a setup applying an agitation speed of 800 rpm and a H_2 gassing rate of 0.0083 vvm. Under the aforementioned conditions a MER of 5.39 CH_4 L^{-1} h^{-1} (calculated from H_2 consumption rate) was achieved. The highest MER of 10.60 CH_4 L^{-1} h^{-1} was achieved by applying an agitation speed of 800 rpm, however at the expense of reduced fermentation offgas quality.

Technical grade H_2/CO_2 conversion was carried out with a mixed microbial hydrogenotrophic and methanogenic enrichment culture in a 5 L stirred tank bioreactor with 2 L working volume at 37 and 60 °C, respectively [28]. Fermentations were performed in continuous culture with and without cell recycle. Also barophilic fermentations were conducted at 37 °C with up to 3109.5 kPa in a 2 L specially constructed bioreactor containing 1 L working volume [28]. Unfortunately, no information on the agitation speed and almost no information on the H_2 gassing rates are provided in the publication. The mixed anaerobic enrichment culture was obtained after 7 days and by applying a HRT of 8.2 days a MER of 24.7 mmol CH_4 L^{-1} h^{-1} was obtained in the 5 L bioreactor. During overpressure experiments at 1385.9 kPa the mixed hydrogenotrophic methanogenic enrichment culture produced CH_4 at a MER of 40.2 mmol CH_4 L^{-1} h^{-1}. However, only after an extended lag time for enrichment culture adaption. In another series of experiments the mixed hydrogenotrophic and methanogenic enrichment culture was successfully adapted to a fermentation temperature of 60 °C and a significant increase of MER could be detected. Although supported by cell recycle, a MER of 446.2 mmol CH_4 L^{-1} h^{-1} could be finally obtained. The results of this publication show that enriched microbial cultures can be applied for H_2/CO_2 conversion and that cell recycle with mixed hydrogenotrophic methanogenic cultures is feasible. The results presented in the publication indicate elevated volumetric CH_4 productivity under thermophilic growth conditions, compared to mesophilic growth conditions with and without cell recycle.

2.3 *Ex situ* Microbiological Biogas Upgrading in a Separate Bioreactor By Contacting H_2, Biogas and a Pure Culture of Hydrogenotrophic Methanogens

Microbiological biogas upgrading was performed by contacting biogas, H_2 and a pure culture of *Methanothermobacter thermoautotrophicum* at 62 °C in pH-controlled fed-batch cultivation mode. The bioreactor specification and configuration was unfortunately not provided in the study [29]. The experimental set-up was used to examine gas liquid mass transfer by application of two different hollow fibre membrane cassettes either operated in parallel counter current flow mode or in orthogonal flow mode for H_2-enriched biogas addition to a suspension containing *M. thermoautotrophicum*. An enrichment of the CH_4 content of biogas from 50–60 to 96 vol% could be achieved. During application of the hollow fibre membrane cassette inside the stirred tank bioreactor under orthogonal flow process mode the molar flux of H_2-enriched biogas into the liquid phase was found to be higher compared to the application of hollow fibre module cassette in counter current flow mode outside of the bioreactor. Unfortunately, no volumetric production rates could be calculated from data presented [29]. The authors state that a MER as high as 10400 mmol CH_4 L^{-1} h^{-1} could be obtained (as calculated from MER of biogas production via microbiological biogas upgrading by a difference of 14 MJ m^{-3}). Furthermore, an almost full conversion of CO_2 with H_2 to a final CH_4 and CO_2 content in the fermentation offgas of 96 and 4 vol%, respectively, could be achieved. Microbiological biogas upgrading by contacting H_2- enriched biogas and a pure culture of *M. thermoautotrophicum* could be successfully performed, and the caloric value in the fermentation offgas had been increased from 21 MJ m^{-3} to about 35 MJ m^{-3}.

Microbiological biogas upgrading by addition of external H_2 was performed in chemostat culture by using a pure culture of *Methanothermobacter marburgensis* operated at 65 °C, applying an agitation speed of 1500 rpm and a pressure of 250 kPa in a 10 L laboratory CSTR containing a working volume of 5 L [22]. A H_2 gassing rate of 0.325 vvm was used in for biogas upgrading. Impure biogas was used in the experiment and could be successfully upgraded by addition of external H_2, but unfortunately no information on volumetric productivity and CH_4 vol% in fermentation offgas is presented in the publication. However, the purpose of the study was to develop a method to be able to quantify volumetric productivities and to detect physiological effects on *M. marburgensis* when exposed to different kind of industrial emission flue gasses, and to show that the pure culture of *M. marburgensis* was not negatively affected by uncharacterized compounds contained in these emission flue gasses. The authors state that MER and CH_4 offgas content remained the same in real gas application compared to reference experiments under equal application of gas to liquid mass transfer conditions. The presented method allows comparing MER obtained from application of emission flue gas to MER obtained from application pure H_2/CO_2 under steady state chemostat culture conditions for analysis and quantification of the volumetric productivity of hydrogenotrophic and methanogenic cultures. Furthermore, it was shown that all factors affecting gas to liquid mass transfer were included in the model.

A postprocessing step for microbiological upgrading of synthetic and industrial biogas was performed by using a pure culture of a *M. thermoautotrophicus* DSM 3590 in a laboratory-bench top bioreactor at a temperature of 60 °C and an agitation speed of 700 rpm in a working volume of 3 L in chemostat culture [30]. H_2 gassing rates of 0.067, 0.133 and 0.533 vvm were applied in synthetic biogas upgrading experiments, and a H_2 gassing rate of 0.067 vvm were used in the experiment for microbiological upgrading of industrial biogas. When introducing the synthetic biogas together with H_2 at gassing rates of H_2 of 0.067 vvm and 0.533 vvm a MER of 23.4 and 50.0 mmol CH_4 L^{-1} h^{-1} were achieved, respectively. The experiments with industrial biogas resulted in a MER of 22.3 mmol CH_4 L^{-1} h^{-1}. Hence, only little difference in MER between application of synthetic and industrial biogas was found and upgrading of CO_2 was shown to be successful. However, bioprocess development work still needs to be performed in order to be able to account for the reduced MER during microbiological biogas upgrading. The reduced MER observed during synthetic and industrial biogas upgrading experiments is due to the CH_4 part present in the biogas, which acts as an inert gas, because CH_4 will be transferred from the gas to the liquid phase alongside H_2 and CO_2 and replaces valuable gaseous substrate which would usually be available to the methanogens.

3 Discussion

The above recapitulation focussing on different aspects of microbiological biogas upgrading experiments, and the extraction of different quantitative data available on the different systems (Table 1), show that only few experimental results have been obtained. In situ microbiological biogas upgrading shows only low volumetric production rates, which are a result of the low applied H_2 gassing rates. Moreover, by analysing offgas composition of in situ microbiologically upgraded biogas it could be revealed that a gas exchange of the fermentation offgas from CO_2 to H_2 was performed. Consequently, the calorimetric value of the processed biogas was only upgraded by addition of H_2. Hence, complete transformation of CO_2 with H_2 to CH_4 could not be obtained by in situ biogas upgrading. This is, a full hydrogenotrophic biomethanation of CO_2 during anaerobic digestion of biomass is yet only feasible if all the H_2 can be quantitatively forced into the liquid phase to be alongside CO_2 in order to be convertible to CH_4 by hydrogenotrophic methanogens.

Table 1 Quantitative data for different hydrogenotrophic methanogen-based microbiological biogas upgrading technologies

Upgrading technology	H_2 gassing rate [vvm]	rpm	Temp. [° C]	Bioprocess mode, comments	Vessel type and working volume	CH_4 offgas [vol%]	MER [mmol L^{-1} h^{-1}]	Reference
In situ	0.0005	100	55	Semi-continuous	4.5 L bioreactor, 3.5 L working volume	65 ± 3.3	0.25[a]	[18]
In situ	0.0012	150	55	Semi-continuous, column diffuser	1 L bottle, 0.6 L working volume	53 ± 3	0.56[a]	[27]
In situ	0.0012	300	55	Semi-continuous, column diffuser	1 L bottle, 0.6 sL working volume	68 ± 2.5	0.66[a]	[27]
In situ	0.0012	150	55	Semi-continuous, ceramic diffuser	1 L bottle, 0.6 L working volume	75 ± 3.4	0.69[a]	[27]
Ex situ, mixed culture	0.0021	500	55	Semi-continuous	1 L bottle, 0.6 L working volume	93.5 ± 4.4	1.35[a]	[7]
Ex situ, mixed culture	0.0042	500	55	Semi-continuous	1 L bottle, 0.6 L working volume	95.4 ± 2.8	2.74[a]	[7]
Ex situ, mixed culture	0.0083	500	55	Semi-continuous	1 L bottle, 0.6 L working volume	89.9 ± 4.1	5.25[a]	[7]
Ex situ, mixed culture	0.0083	800	55	Semi-continuous	1 L bottle, 0.6 L working volume	94.2 ± 2.8	5.39[a]	[7]
Ex situ, mixed culture	0.0167	800	55	Semi-continuous	1 L bottle, 0.6 L working volume	90.8 ± 2.8	10.59[a]	[7]
Ex situ, mixed culture	n.a.	n.a.	60	Continuous culture	n.a	n.a	258.77[a]	[28]
Ex situ, mixed culture	n.a.	n.a.	60	Continuous culture, with cell recycle	n.a	n.a	446.15[a]	[28]
Ex situ, mixed culture	n.a.	n.a.	37	Continuous culture	n.a	n.a	24.75[a]	[28]

(continued)

Table 1 (continued)

Upgrading technology	H_2 gassing rate [vvm]	rpm	Temp. [° C]	Bioprocess mode, comments	Vessel type and working volume	CH_4 offgas [vol%]	MER [mmol L^{-1} h^{-1}]	Reference
Ex situ, mixed culture	n.a.	n.a.	37	Continuous culture, with overpressure	n.a	n.a	40.15[a]	[28]
Ex situ, pure culture	n.a.	n.a.	62	Fed-batch	n.a	96	10400[b]	[29]
Ex situ, pure culture	0.325	1500	65	Chemostat culture, overpressure	10 L bioreactor, 5 L working volume	n.a.	n.a.	[22]
Ex situ, pure culture	0.067	700	60	Chemostat culture	bioreactor, 3 L working volume	n.a.	23.42[c]	[30]
Ex situ, pure culture	0.533	700	60	Chemostat culture	Bioreactor, 3 L working volume	n.a.	50.01[c]	[30]
Ex situ, pure culture	0.067	700	60	Chemostat culture	Bioreactor, 3 L working volume	n.a.	22.31[c]	[30]

n.a. not attainable
[a]Calculated from volumetric H_2 uptake rate divided by four
[b]Calculated from MER of biogas production (microbiological biogas upgrading by a value of 14 MJ m^{-3})
[c]Calculated from volumetric CH_4 production rate

However, an increase of the H_2 gassing rate, combined with an increase of agitation speed as well as an increase of cultivation pressure, would consequently result in an elevated H_2 gas to liquid mass transfer, which is expected to result in higher MER values during in situ microbiological biogas upgrading.

Another unresolved problem for upgrading biogas by contacting H_2 and anaerobic digester sludge inside fermentors emerges from underestimated concentrations of H_2 residing in bubbles inside anaerobic digesters [31]. Poorly soluble gasses (i.e. H_2) will remain overconcentrated inside ascent bubbles. This is, a higher amount of dissolved H_2, than measured during anaerobic digestion fermentation offgas analysis, is already present in the liquid phase of the highly viscous anaerobic sludge due to limited mixing [31]. However, the behaviour of additionally added H_2 during in situ microbiological biogas upgrading is difficult to assess in relation to anaerobic digestion, and the possible discussion remains to be resumed, and results are to be re-interpreted, when new data has been collected upon H_2 concentration probing from inside the anaerobic digester sludge (during in situ microbiological biogas upgrading experiments).

In situ microbiological biogas upgrading was until now only performed in laboratory type anaerobic digesters. Development of the technology regarding volumetric productivity and complete conversion to biomethane is still on demand. Thereafter, and if proper H_2 gas to liquid mass transfer can be achieved, the technology readiness has to be proven in pilot plant scale. However, the below raised questions regarding in situ microbiological biogas upgrading technology remain not the only issues to be addressed before an industrial application can be performed:

- How do irregular periods of H_2 introduction (as well as subsequent changes of pH, oxidation-reduction potential (ORP), CO_2 removal etc.) affect the microbial population structure, productivity and fermentation offgas composition?
- How will the in situ microbiological biogas upgrading technology be integrated to serve as power-to-gas conversion technology when irregular supply of H_2 occur, and an immediate start-up (or shut-down) of the in situ upgrading system would be urgently required?
- How can the feeding ratio of H_2 to CO_2 being controlled online at 4:1 in order to be able to quantitatively convert all CO_2 into CH_4?
- How can the introduction of H_2 into industrial scale type anaerobic digesters being accomplished regarding engineering, infrastructure investments, upstream and downstream processing, energy return of energy investment, etc.?
- What is needed to re-adapt existing security measurements in order to be able to meet commercialization of the in situ microbiological biogas upgrading technology, and where is an installation of such a technology useful and appropriate?

Applying the anaerobic digestion model No. 1 (ADM1) could answer some of the above raised questions. In a recent study published by Bensmann et al. the authors adapted ADM1 to the extent that in situ microbiological biogas upgrading has been included in the simulation [32]. The authors modelled different scenarios in which either biological limitation or H_2 gas to liquid mass transfer limitation was included in the model. An emphasis was given to model the influence of the specific transport coefficient of H_2 (k_La_{H2}) on in situ microbiological biogas upgrading. Furthermore, the authors simulated whether control strategies could be included in the modified ADM1 version. The authors suggest a scenario in which specific H_2 transfer coefficients k_La_{H2} of 30 h^{-1} should be applied. However, there is only a restricted number of publications available on k_La_{H2} values that are commonly found in anaerobic digesters. This is, typical k_La_{H2} values are in the range of approximately 9 h^{-1} [33]. Hence, as suggested before, more quantitative data is needed for in situ microbiological biogas upgrading, not only from simulations, but also from experiments.

Ex situ conversion of a synthetic biogas mixture or technical grade H_2/CO_2 by enriched hydrogenotrophic methanogenic consortia was performed in separate bioreactors during a postprocessing step in semi-continuous [7] and continuous culture [28], respectively. Comparing MER values, presented from mesophilic versus thermophilic culture conditions for mixed hydrogenotrophic methanogens used for CH_4 production (Table 1), reveals an increase of MER by >90 % due to application of thermophilic cultivation conditions [28]. However, this was only the result of a single analysis and more experimental data is obviously required to support this conclusion. In this respect, one has to also keep in mind that enriching hydrogenotrophic methanogenic microbial cultures might require a pre-adaption period in order to make the culture fully acclimated to special fermentation temperatures and/or H_2/CO_2 gassing rates, a *modus operandi* which is not required for pure culture of microorganisms. The results of above presented studies also show that a fully acclimated hydrogenotrophic methanogenic enrichment culture can sustain overpressure conditions and is furthermore capable of microbiological biogas upgrading as well as of achieving high MERs [28], a finding, in controversy to discussions found elsewhere [30]. Eventually, during experiments applying a monoculture of hydrogenotrophic methanogens it has been shown, that an increase of the pressure inside the bioreactor will elevate the MER, which is a consequence of increasing the maximum solubility of gasses inside the liquid phase [23, 24, 30]. Hence, by applying overpressure an increase of the MER could be achieved, if a hydrogenotrophic and methanogenic enrichment culture would be utilized for ex situ microbiological biogas upgrading.

Interestingly, also a cell-recycle system was used to increase the MER during H_2/CO_2 conversion experiment using a mixed microbial culture, compared to only low MERs which have been experimentally determined when semi-continuous culture or continuous culture had been applied for microbiological biogas upgrading [28]. This finding is consistent to results obtained for a pure culture of *M. marburgensis* [23]. During application of the cell-recycle system the MER inceased by 42 % compared to standard bioprocessing conditions [28]. In contrast to the utilization of the in situ microbiological biogas upgrading technology the H_2 gas to liquid mass transfer is much higher for suspensions containing pure or enrichment cultures of hydrogenotrophic methanogens, because the viscosity of such a free floating cell suspension is lower compared to the highly viscous anaerobic digester sludge.

Another issue concerning the stability of the enrichment culture could arise from application of dynamic process conditions. This is, a technology applied for microbiological biogas upgrading would be required starting up very fast, if urgent storage of renewable generated electricity via electrolysis of H_2O to H_2 and conversion of H_2 with CO_2 to CH_4 is on demand. How can the stability of a mixed hydrogenotrophic culture be assured during off times? How long does it take to start up a bioreactor containing a non-pre adapted mixed microbial culture to be enriched for a specific gas mixture? However, mixed microbial cultures might nevertheless have advantages compared to pure culture of microorganisms in a scenario wherein not only the CO_2 of biogas is upgraded to biomethane, but in which also CO_2 or H_2-containing industrial emission flue gasses have to be converted. An isolation of novel hydrogenotrophic methanogenic strains from enrichment cultures, exclusively adapted towards gas composition fluctuations, gassing rates, temperature as well as to other environmental challenging conditions has yet not been attempted, but could result in the identification and characterisation of useful novel strains for pure culture H_2/CO_2 to CH_4 conversion. Furthermore, ex situ microbiological biogas upgrading by contacting biogas, H_2 and pure or enrichment cultures of hydrogenotrophic methanoges in a separate bioreactor has the advantage of applying individual bioprocess control strategies, applying sophisticated bioreactor instrumentation as well as streamlining the upstream and downstream processing area.

Yet, ex situ microbiological biogas upgrading by contacting biogas, H_2 and pure culture of hydrogenotrophic methanogens did also not receive much attention in the scientific community [22, 29, 30]. In all three available publications concerning pure culture microbiological biogas upgrading only different thermophilic hydrogenotrophic methanogenic strains were used for bioprocessing. This is an interesting finding, because H_2 solubility is lower at higher temperatures, but likely the advantages of higher conversion speed and reduced contamination risk at elevated temperatures are strong arguments in favour of using thermophilic hydrogenotrophic methanogens for microbiological biogas upgrading. The utilization of thermophilic hydrogenotrophic methanogens in biogas upgrading could be also just have performed arbitrarily, because none of the authors does explicitly state why especially thermophilic hydrogenotrophic methanogens would be especially suitable for microbiological biogas upgrading. However, these results are in good agreement to the findings made for pure H_2/CO_2 gas conversion using pure culture of hydrogenotrophic methanogens, wherein almost exclusively *Methanothermobacter* spp. were used for bioprocess development [9]. One reason could be that a non-cytochrome containing microbe could be the right choice to set-up a biological methanation process, because it shows a lower half-saturation constant for the uptake of H_2. Nevertheless, it remains to be shown, why just *Methanothermobacter* spp. should be the right choice to be applied in pure culture bioprocessing for microbiological biogas upgrading?

The bioprocess operation modes used for examination of microbiological biogas upgrading were fed-batch [29] and chemostat culture [22, 30]. Especially cultivation in chemostat culture mode allows structured bioprocess optimization and development, which is required in order to overcome the limitations arising during conversion of gas mixtures containing inert gas compounds [22]. Nevertheless, experiments in chemostat culture are important, because a continuous culture allows examining one of the main industrial prerequisites, which is long-term process stability. Moreover, productivity and other relevant process parameters can be easily compared in chemostat culture, because conditions remain stable for a certain period of time as defined by the operator. However, during chemostat culture operation biomass is washed out of the bioreactor and therefore possibilities circumventing biomass loss need to be established, especially in anaerobic cultures which are usually limited in biomass productivity. Furthermore, methods and techniques other than chemostat culture bioprocessing have to be developed and implemented for holistic microbiological biogas upgrading, which could allow comparing productivity and physiology among different hydrogenotrophic methanogens in a much faster way.

From a methodological point of view the characterisation of novel hydrogenotrophic methanogens from pure or enrichment cultures for efficient microbiological biogas upgrading is highly on demand, because bioprocess development for biological methanation was, as reviewed elsewhere [9], yet mainly excessively performed by using invariant experiments in fed-batch or chemostat culture fermentations. Hence, bioprocess development could be improved by using e.g. dynamic process conditions or by using Design of Experiments (DoE) examination strategies for rapid screening of relevant process parameters. The implementation and the use of DoE for bioprocess development and the utilisation of dynamic process conditions is advantageous, because it accelerates research, if data exploitation can be properly performed [34]. By using dynamic process conditions (pulse, shift, ramps or oscillations), knowledge of maximum biological capacity [21] or yields can be investigated in a short time. Also the examination of limitations, changing productivities, yields, and metabolic states can be achieved through dynamic experiments [34]. Furthermore by precisely altering process conditions through application of dynamic experiments putative liquid and gaseous limitations can be rapidly detected and physiological parameters, which are important for scale-up and bioprocess development, may be rapidly determined. A very good example for bioprocess development applying a combination of multivariate and dynamic experiments using a pure culture of methanogens has only recently presented [35]. Useful results could be gained from multivariate and dynamic experiments in short time and novel physiological responses interlinked with new findings on the productivity of *M. marburgensis* could be revealed.

This comprehensive literature analysis of in situ and ex situ microbiological biogas upgrading technologies by using hydrogenotrophic and methanogenic enrichment cultures revealed that no targeted bioprocess development has yet been performed. However, many publications on bioprocess development using hydrogenotrophic methanogens grown on pure H_2/CO_2 are already available [9, 21–23, 35], which could speed up bioprocess development of the microbiological biogas upgrading technology in respect to the utilization of mixed cultures of hydrogenotrophic methanogens. Regarding in situ microbiological biogas upgrading aforementioned queries concerning bioprocess development need to be overcome before even pilot plant scale could be implemented. However, this is not the case for ex situ microbiological biogas upgrading technologies using pure culture of hydrogenotrophic methanogens whereof bioprocess development was already finished in lab scale conditions and pilot plants are currently being considered, started-up or are already under operation.

Moreover, a pure or a mixed culture of hydrogenotrophic methanogens could be used for upgrading CO_2, emitted by other biofuel production processes (biohydrogen or bioethanol) [17, 36–40], or from wastewater treatment plants, by addition of H_2. Such a broad applicability and flexibility of hydrogenotrophic methanogens for conversion of H_2 and CO_2 containing emission waste gasses was already shown for pure culture of methanogens [22]. Hence, it can be assumed, that also hydrogenotrophic methanogenic enrichment cultures are similarly capable of converting H_2/CO_2-containing impure reactant gasses—something currently not being achievable with a chemical catalyst. On the contrary, anaerobic digesters are a source of biological CO_2 emission and, hence, the question might be asked whether an anaerobic digester could provide additional CO_2 uptake and conversion capacity and to what extent such a modification of an anaerobic digester is feasible and useful?

Major concerns for successful bioprocess development for microbiological biogas upgrading by addition of H_2 to pure or mixed consortia of microbes in a separate bioreactor includes the following issue: CH_4 in biogas will act as an inert gas. This is, CH_4 is transferred alongside with CO_2 and H_2 from the gas into the liquid phase, reducing the gas to liquid mass transfer of H_2/CO_2 [22, 30], which was found to reduce the volumetric CH_4 productivity tremendously [30]. In this respect, bioreactors other than stirred tank reactors could be considered [9], [41, 42]. Furthermore, high volumetric CH_4 productivities are usually demanded for industrial processes. High volumetric CH_4 productivity can be achieved by increasing the H_2/CO_2 gassing rate, but then a lower offgas product quality will be observed [23], [30]. However, in this relation also the application of overpressure conditions needs to be considered [23, 24, 30]. Eventually, in order to achieve a full conversion of CO_2 with H_2 a ratio of 1:4 has to be applied [9, 23], otherwise excess H_2 or CO_2 will be remaining in the fermentation offgas.

4 Conclusions

Microbiological biogas upgrading for carbon capture and utilization using hydrogenotrophic methanogenic Archaea seems to be a very promising technology for storage of electricity generated during peak times via electrolysis of H_2O to H_2 and subsequent reduction of CO_2 to CH_4.

The conversion and microbiological upgrading of CO_2 to biomethane is feasible. The quality of biomethane was found to exceed or is close to 95 vol%. At such offgas quality the biomethane could be introduced into the natural gas grid (depending on national regulations).

The production of CO_2 emitted after combusting biomethane would be carbon neutral, because the released CO_2 had initially been incorporated into biomass by photosynthetic organisms.

In situ microbiological biogas upgrading in anaerobic digesters still requires thorough bioprocess development. Otherwise this process will remain at a low technology development stage compared to ex situ microbiological biogas upgrading applying pure or mixed culture of hydrogenotrophic methanogens.

Ex situ microbiological biogas upgrading mainly suffers from inert CH_4 gas liquid mass transfer into the liquid phase during microbiological biogas upgrading, which results in low volumetric CH_4 productivity compared to pure H_2/CO_2 conversion.

Quantitative analysis of the microbiological biogas upgrading process regarding commercialization, life cycle assessment and the energy return of energy investment needs to be performed in order to be able to adequately compare the different technologies.

If waste gas (originating from biological or industrial sources) utilization can be shown with only little reduction in volumetric CH_4 productivity, as well as if the aim of a single stage conversion to a CH_4 fermentation offgas content exceeding 95 vol% can be demonstrated, ex situ microbiological biogas upgrading with pure or mixed cultures could become a promising future technology for almost CO_2-neutral renewable natural gas production.

Acknowledgments Ruth-Sophie Taubner and Günther Bochmann are gratefully acknowledged for proof reading and critical comments on the manuscript.

References

1. Solomon S, Plattner G-K, Knutti R, Friedlingstein P (2009) Irreversible climate change due to carbon dioxide emissions. Proc Natl Acad Sci 106(6):1704–1709
2. Howarth RW (2014) A bridge to nowhere: methane emissions and the greenhouse gas footprint of natural gas. Energy Sci Eng 2(2):47–60
3. Marshall J, Armour KC, Scott JR, Kostov Y, Hausmann U, Ferreira D, Shepherd TG, Bitz CM (2014) The ocean's role in polar climate change: asymmetric Arctic and Antarctic responses to greenhouse gas and ozone forcing. Philos Trans R Soc Math Phys Eng Sci 372 (2019):20130040
4. Siegl S, Laaber M, Holubar P (2011) Green electricity from Biomass, part I: environmental impacts of direct life cycle emissions. Waste Biomass Valorization 2(3):267–284
5. Siegl S, Laaber M, Holubar P (2012) Green electricity from biomass, part II: environmental impacts considering avoided burdens from replacing the conventional provision of additional functions. Waste Biomass Valorization 3(1):1–21
6. Ryckebosch E, Drouillon M, Vervaeren H (2011) Techniques for transformation of biogas to biomethane. Biomass Bioenergy 35(5):1633–1645
7. Luo G, Angelidaki I (2012) Integrated biogas upgrading and hydrogen utilization in an anaerobic reactor containing enriched hydrogenotrophic methanogenic culture. Biotechnol Bioeng 109(11):2729–2736

8. Yang L, Ge X, Wan C, Yu F, Li Y (2014) Progress and perspectives in converting biogas to transportation fuels. Renew Sustain Energy Rev 40:1133–1152
9. Rittmann S, Seifert A, Herwig C (2015) Essential prerequisites for successful bioprocess development of biological CH_4 production from CO_2 and H_2. Crit Rev Biotechnol 35(2):141–151
10. Bernacchi S, Seifert AH, Krajete A, Rittmann SK-MR (2013) Anwendungen der Methanogenese zur Biogasveredelung und Stromspeicherung presented at the biogas 13. St. Pölten, Austria, 04 Dec 2013
11. Rittmann SK-MR, Seifert AH, Krajete A (2014) Biomethanisierung—ein Prozess zur Ermöglichung der Energiewende? BIOspektrum 20(7):816–817
12. Offre P, Spang A, Schleper C (2013) Archaea in biogeochemical cycles. Annu Rev Microbiol 67(1):437–457
13. Cavicchioli R (2011) Archaea–timeline of the third domain. Nat Rev Microbiol 9(1):51–61
14. Thauer RK, Kaster A-K, Seedorf H, Buckel W, Hedderich R (2008) Methanogenic archaea: ecologically relevant differences in energy conservation. Nat Rev Microbiol 6(8):579–591
15. Thauer RK, Kaster A-K, Goenrich M, Schick M, Hiromoto T, Shima S (2010) Hydrogenases from methanogenic archaea, nickel, a novel cofactor, and H_2 storage. Annu Rev Biochem 79:507–536
16. Schönheit P, Moll J, Thauer RK (1980) Growth parameters (K s, μmax, Y s) of Methanobacterium thermoautotrophicum. Arch Microbiol 127(1):59–65
17. Rittmann SK-MR, Lee HS, Lim JK, Kim TW, Lee J-H, Kang SG (2015) One-carbon substrate-based biohydrogen production: Microbes, mechanism, and productivity. Biotechnol Adv 33(1):165–177
18. Rittmann S, Holubar P (2014) Rapid extraction of total RNA from an anaerobic sludge biocoenosis. Folia Microbiol (Praha) 59(2):127–132
19. Weiland P (2003) Production and energetic use of biogas from energy crops and wastes in Germany. Appl Biochem Biotechnol 109(1–3):263–274
20. Liu Y, Whitman WB (2008) Metabolic, phylogenetic, and ecological diversity of the methanogenic archaea. Ann N Y Acad Sci 1125:171–189
21. Rittmann S, Seifert A, Herwig C (2012) Quantitative analysis of media dilution rate effects on Methanothermobacter marburgensis grown in continuous culture on H_2 and CO_2. Biomass Bioenergy 36:293–301
22. Seifert AH, Rittmann S, Bernacchi S, Herwig C (2013) Method for assessing the impact of emission gasses on physiology and productivity in biological methanogenesis. Bioresour Technol 136:747–751
23. Seifert AH, Rittmann S, Herwig C (2014) Analysis of process related factors to increase volumetric productivity and quality of biomethane with Methanothermobacter marburgensis. Appl Energy 132:155–162
24. Nishimura N, Kitaura S, Mimura A, Takahara Y (1992) Cultivation of thermophilic methanogen KN-15 on H2-CO2 under pressurized conditions. J Ferment Bioeng 73(6): 477–480
25. Luo G, Johansson S, Boe K, Xie L, Zhou Q, Angelidaki I (2012) Simultaneous hydrogen utilization and in situ biogas upgrading in an anaerobic reactor. Biotechnol Bioeng 109 (4):1088–1094
26. Drake HL, Daniel SL, Küsel K, Matthies C, Kuhner C, Braus-Stromeyer S (1997) Acetogenic bacteria: what are the in situ consequences of their diverse metabolic versatilities? BioFactors Oxf Engl 6(1):13–24
27. Luo G, Angelidaki I (2013) Co-digestion of manure and whey for in situ biogas upgrading by the addition of H2: process performance and microbial insights. Appl Microbiol Biotechnol 97 (3):1373–1381
28. Wise DL, Cooney CL, Augenstein DC (1978) Biomethanation: Anaerobic fermentation of CO_2, H_2 and CO to methane. Biotechnol Bioeng 20(8):1153–1172
29. Strevett KA, Vieth RF, Grasso D (1995) Chemo-autotrophic biogas purification for methane enrichment: mechanism and kinetics. Chem Eng J Biochem Eng J 58(1):71–79

30. Martin MR, Fornero JJ, Stark R, Mets L, Angenent LT (2013) A single-culture bioprocess of Methanothermobacter thermautotrophicus to upgrade digester biogas by CO_2-to-CH_4 by conversion with H_2. Archaea 2013:e157529
31. Pauss A, Andre G, Perrier M, Guiot SR (1990) Liquid-to-gas mass transfer in anaerobic processes: inevitable transfer limitations of methane and hydrogen in the biomethanation process. Appl Environ Microbiol 56(6):1636–1644
32. Bensmann A, Hanke-Rauschenbach R, Heyer R, Kohrs F, Benndorf D, Reichl U, Sundmacher K (2104) Biological methanation of hydrogen within biogas plants: A model-based feasibility study. Appl Energy 134:413–425
33. Sonesson M (2013) Methane yields from anaerobic digestion of food waste. Karlstad University, Sweden
34. Spadiut O, Rittmann S, Dietzsch C, Herwig C (2013) Dynamic process conditions in bioprocess development. Eng Life Sci 13(1):88–101
35. Bernacchi S, Rittmann S, Seifert AH, Krajete A, Herwig C (2014) Experimental methods for screening parameters influencing the growth to product yield (Y(x/CH4)) of a biological methane production (BMP) process performed with Methanothermobacter marburgensis. AIMS Bioeng. 1(2):72–86
36. Martinez-Porqueras E, Rittmann S, Herwig C (2013) Analysis of H_2 to CO_2 yield and physiological key parameters of Enterobacter aerogenes and Caldicellulosiruptor saccharolyticus. Int J Hydrog Energy 38(25):10245–10251
37. Martinez-Porqueras E, Rittmann S, Herwig C (2012) Biofuels and CO_2 neutrality: an opportunity. Biofuels 3(4):413–426
38. Rittmann S, Herwig C (2012) A comprehensive and quantitative review of dark fermentative biohydrogen production. Microb Cell Fact 11(1):115
39. Ahring BK, Westermann P (2007) Coproduction of bioethanol with other biofuels. Adv Biochem Eng Biotechnol 108:289–302
40. Antoni D, Zverlov VV, Schwarz WH (2007) Biofuels from microbes. Appl Microbiol Biotechnol 77(1):23–35
41. Burkhardt M, Busch G (2013) Methanation of hydrogen and carbon dioxide. Appl Energy 111:74–79
42. Lee JC, Kim JH, Chang WS, Pak D (2012) Biological conversion of CO_2 to CH_4 using hydrogenotrophic methanogen in a fixed bed reactor. J Chem Technol Biotechnol 87(6): 844–847

Influent Fractionation for Modeling Continuous Anaerobic Digestion Processes

Manfred Lübken, Pascal Kosse, Konrad Koch, Tito Gehring and Marc Wichern

Abstract The first dynamic model developed to describe anaerobic digestion processes dates back to 1969. Since then, considerable improvements in identifying the underlying biochemical processes and associated microorganisms have been achieved. These have led to an increasing complexity of both model structure and the standard set of stoichiometric and kinetic parameters. Literature has always paid attention to kinetic parameter estimation, as this determines model accuracy with respect to predicting the dynamic behavior of biogas systems. As sufficient computing power is easily available nowadays, sophisticated linear and nonlinear parameter estimation techniques are applied to evaluate parameter uncertainty. However, the uncertainty of influent fractionation in these parameter optimization procedures is generally neglected. As anaerobic digestion systems are currently increasingly used to convert a broad variety of organic biomass to methane, the lack of generally accepted guidelines for input characterization adapted to the simulation model's characteristics is a considerable limitation of model application to these substrates. Directly after the introduction of the standardized Anaerobic Digestion Model No. 1 (ADM1), several publications pointed out that the model's requirement of a detailed influent characterization can hardly be fulfilled. The main shortcoming of the model application was addressed in the reliable and practical identification of the model's input state variables for particulate and soluble carbohydrates, proteins and lipids, as well as for the inerts. Several authors derived biomass characterization procedures, most of them dedicated to a particular substrate, and some of them being of general nature, but none of these approaches have resulted in a practical standard protocol so far. This review provides an overview of

M. Lübken (✉) · P. Kosse · T. Gehring · M. Wichern
Institute of Urban Water Management and Environmental Engineering,
Ruhr-Universität Bochum, Bochum, Germany
e-mail: manfred.luebken@ruhr-uni-bochum.de

K. Koch
Chair of Urban Water Systems Engineering, Technische Universität München,
Munich, Germany
e-mail: k.koch@tum.de

G.M. Guebitz et al. (eds.), *Biogas Science and Technology*,
Advances in Biochemical Engineering/Biotechnology 151,
DOI 10.1007/978-3-319-21993-6_6

existing approaches that improve substrate influent characterization to be used for state of the art anaerobic digestion models.

Keywords Anaerobic digestion · Mathematical modeling · ADM1 · Influent characterization · Fractionation

Contents

1 Introduction

Anaerobic digestion (AD) is a multistep process and its complexity has always stimulated the development of mathematical models to understand the manifold interactions of substances and microorganisms assigned to each other. Andrews [1] developed the first AD model in 1969. It was based on the rate-limiting step approach and considered only the acetoclastic methanogenic process step. The list of models following the pioneering work of Andrews is extensive as model development was not directed and a standardized basis was not available. Several review works have been produced so far to sort the publications within a specific direction. Husain [2], for instance, reviewed both steady state and dynamic models able to accurately describe the kinetics of anaerobic digestion, Saravanan and Sreekrishnan [3] analyzed suitable models for anaerobic biofilm reactors and Donoso-Bravo et al. [4] addressed methodologies and achievements in parameter estimation and identification. Substrate specific reviews are available for sewage sludge [5], lignocellulosic biomass [6] and domestic wastewater [7].

The generic Anaerobic Digestion Model No. 1 (ADM1) is currently the most frequently used approach for modeling AD of complex wastes, focusing on the principle processes known as disintegration, hydrolysis, acidogenesis (or fermentation), acetogenesis, and methanogenesis [8, 9]. The term "complex" here refers to the degradation pathways of the prime products: soluble and particulate carbohydrates (X_{ch}), proteins (X_{pr}), lipids (X_{li}) and volatile fatty acids (VFAs). As organic wastes are very heterogeneous by nature [10, 11], the key to successful modeling using the ADM1 is subject to detailed knowledge of the substrate influent characteristics, because such fractions trigger the anaerobic degradability.

The ADM1 was mainly designed to be readily extendible and, thus, various extensions followed, focusing, for instance, on modeling biodegradation processes of phenol compounds [12] and on sulfate reduction [13]. However, in addition to extension development, scientists also criticized constructively in order to encourage improvement of the ADM1. Probably the most serious criticism lies in the substrate characterization that is closely connected to the hydrolysis step for which the rates differ for particulate components, such as carbohydrates, proteins and lipids [14, 15]. Generally, ADM1 describes the degradation of composite particulate material (X_c) with lumped characteristics (see Fig. 1), while the hydrolysis steps describe well-defined, relatively pure substrates, such as carbohydrates, lipids and proteins.

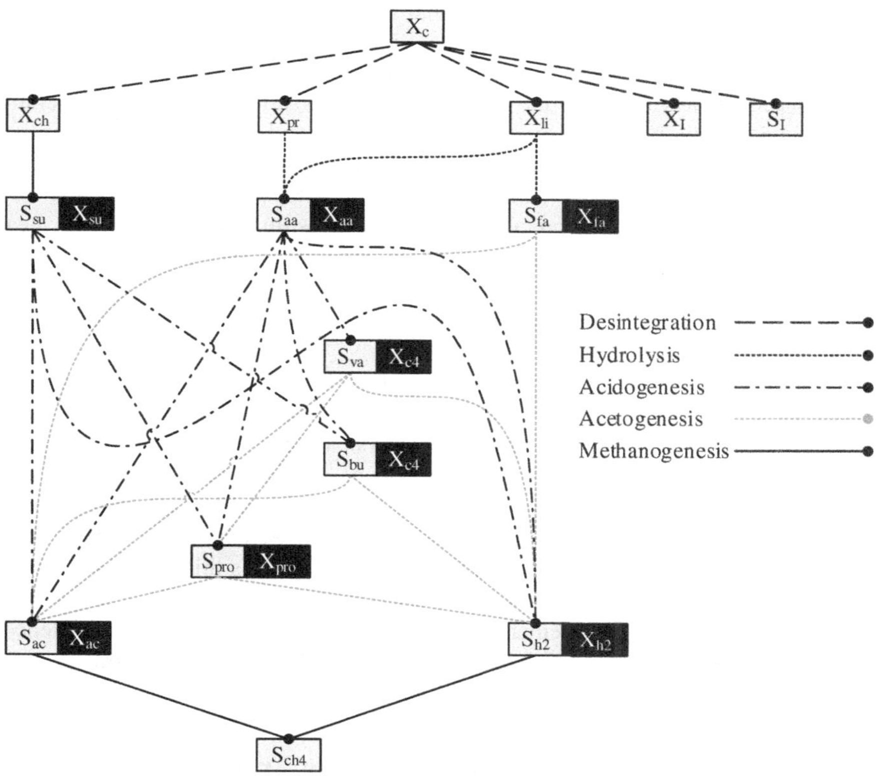

Fig. 1 Basic reaction scheme of ADM1 for organic components. X_c composite material, X_{ch} particulate carbohydrates, X_{pr} particulate proteins, X_{li} lipids, X_I particulate inerts, S_I soluble inerts, S_{su} monosaccharides, S_{aa} amino acids, S_{fa} long chain fatty acids, S_{va} valerate, S_{bu} butyrate, S_{pro} propionate, S_{ac} acetate, S_{h2} hydrogen, S_{ch4} methane, X_{su} sugar degrader, X_{aa} amino acids degrader, X_{fa} long chain fatty acids degrader, X_{c4} valerate and butyrate degrader, X_{pro} propionate degrader, X_{ac} acetate degrader and X_{h2} hydrogen consumer. Basic unit of organic components is $kg_{COD} \cdot m^{-3}$

All those organic species, as well as molecular hydrogen, are described in terms of the chemical oxygen demand (COD), while inorganic carbon or inorganic nitrogen species are described on a molar basis. Due to the heterogeneous nature of organic wastes, it is neither possible to fix the waste to one lumped characteristic nor practical to characterize it as a pure substrate [16–18].

The COD of any digester input can generally be divided into two main fractions, the biodegradable and unbiodegradable part. The unbiodegradable part is composed of two sub-fractions, particulate inerts (X_I) and soluble inerts (S_I). The biodegradable part is handled in a rather complex manner within the ADM1 framework and is composed of four particulates: X_c, X_{ch}, X_{pr} and X_{li}, as well as seven soluble state variables: S_{su}, S_{aa}, S_{fa}, S_{bu}, S_{va}, S_{pro} and S_{ac}. The fraction X_c represents the precursor when biomass with lumped characteristics, such as sewage sludge, is treated. However, as pointed out in [19], the default X_c composition in terms of carbohydrates, proteins, lipids and inert fractions is not in agreement with the typical composition of sewage sludge. A particular disagreement was found for the particulate inert fraction, which was set to 44 % of X_C, in contrast to the default value of 25 %. Nevertheless, a particular challenge in the correct determination of the single X_C fractions always exists in the fact that the composition of biomass generally varies significantly, both in terms of biodegradability and degradation kinetics.

Dynamic online monitoring systems have proven to be applicable in AD systems [20–22], but these systems are generally used to monitor process stability and they allow only a limited range of process variables to be estimated. More recently, near infrared spectroscopy (NIRS) was successfully applied for the rapid determination of feedstock biodegradability [23, 24]. These methods would allow a dynamic characterization of the digester influent, but they have not yet been coupled to determine ADM1 input state variables.

Various studies have already been devoted to the issue of substrate characterization in ADM1 [16, 17, 19, 25–33]. The methods for the determination of ADM1 input state variables can be grouped into physical-chemical analysis, elemental analysis, anaerobic respirometry, combination of different model interfaces and physical-chemical analysis combined with online gas curve calibration procedure [16, 30]. This review discusses the most relevant approaches which have been elaborated in recent literature. The focus is on the ADM1, as this model represents the leading current modeling framework for simulating AD processes, and this model has the highest requirements for a detailed influent fractionation. Hence, most of the approaches can be reduced to more simple modeling frameworks.

2 Determination of the Unbiodegradable Fraction

In very general terms, biomass can be split into three main components (Fig. 2):

Water

Organics (commonly qualified as volatile solids (VS) or organic dry matter (oDM)

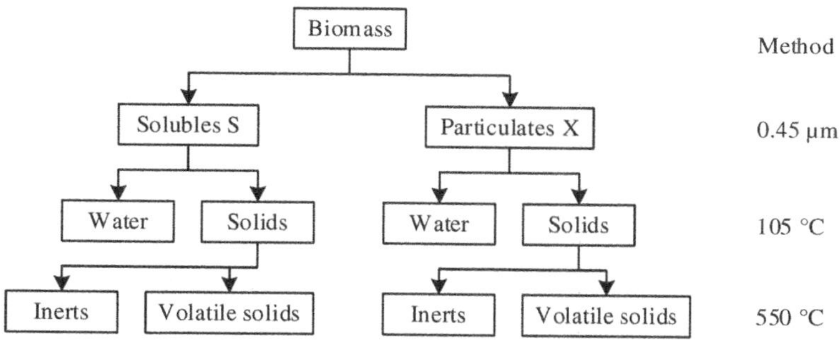

Fig. 2 First-step characterization of input biomass and the method of determination

Inerts (inorganics/ash)

Quantification of the inert and, hence, unbiodegradable fraction is the first step in estimating the biodegradability of the input [34]. In ADM1, the factor D was introduced to describe the ultimate biodegradability, where D = 1 indicates a totally biodegradable organic compound, which is seldom found. Furthermore, the authors of the ADM1 suggest using the term "substrate" to indicate biodegradable organics (D), while unbiodegradable organics (1-D) are termed inerts. In order to avoid confusion with the input biomass, which is also often termed substrate, the organic fraction of both biodegradable and unbiodegradable is simply summarized as organics.

Models, such as the ADM1, typically distinguish between soluble (S) and particulate (X) material. In order to obtain the two fractions, the input is usually prepared through a filter with a 0.45 µm pore size [35–38]. Afterwards, the two fractions are dried at 105 °C (to obtain the amount of total solids: TS) and thereafter at 550 °C (to obtain the amount of VS) until a constant weight is reached [39, 40]. However, the loss of volatile compounds during TS and VS determination (such as VFAs, lactic acid and alcohols) should be taken into account, especially when ensiled energy crops are used as input biomass, and hence, correction should be applied [41]. Figure 2 summarizes a first step for the characterization of input biomass and the method of determination, as mentioned above.

As a basis for more complex models, lumped parameters are chosen to account for a wide range of substances. The most common parameters are the COD for rather liquid biomass, such as sewage sludge, and the VS concentration for rather solid biomass, such as biowaste, manure and energy crops. Both parameters are linked and can be converted by theoretical oxygen demand (ThOD) or by means of empirical approaches (see Chap. 3.1). When applying the approach presented in Fig. 2, the two fractions of soluble (S_I) and particulate (X_I) inerts will be zero, since the method of sample treatment (loss-on-ignition method at 550 °C) causes volatilization of all oxidizable compounds. Hence, this approach is not suitable for the characterization of the inerts at COD or VS base, but demonstrates that inerts in the model's sense are not equal to inerts in the typical analytical sense. Inerts in the model's sense contain COD; this does not mean the few inorganic substances that contribute to COD, such as nitrite and iodide, but unanaerobically or hardly anaerobically degradable organic

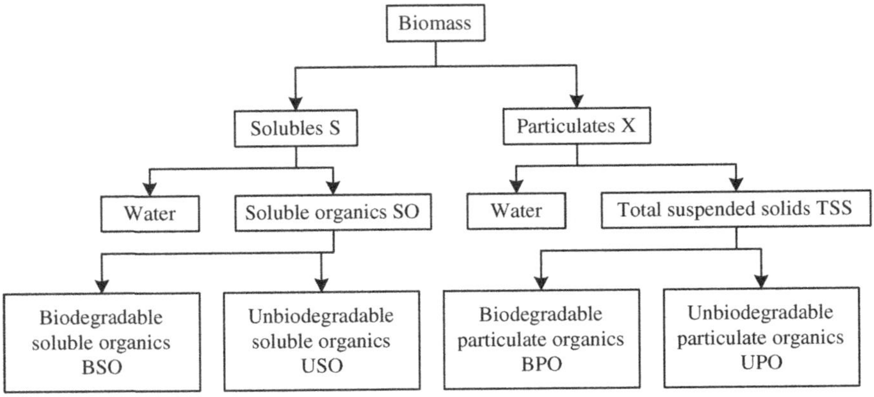

Fig. 3 Characterization of input biomass, modified according to [42]

substances, such as lignin and cellulose/hemicellulose [33]. Figure 3 depicts a possible characterization of the input biomass according to an approach suggested in [42], in which unbiodegradable soluble organics correspond to S_I and unbiodegradable particulate organics to X_I in the ADM1. However, analysis of the several fractions is not as simple as in the first-step in Fig. 2 and will, therefore, be discussed in the following subchapters.

2.1 Determination of the Soluble Inert Fraction S_I

The influent COD for typical input biomass, such as sewage sludge, biowaste, manure or energy crops, is clearly dominated by the particulate fraction. The soluble fraction in general and the soluble inert fraction in particular often play only a minor role in the influent characterization. However, their quantification is important in order to assess the efficiency of the overall AD process, knowing that 350 mL methane is formed when 1 g of COD is degraded [43]. The most popular method to determine the biodegradability D and, hence, the soluble inert fraction (1-D), are biochemical methane potential (BMP) measurement tests with the filtered sample [44]. Therefore, the cumulated methane production (NmL_{CH4}) is normalized with the COD mass of the substrate introduced (g_{COD}). Biodegradability D is obtained by dividing the value by the theoretical BMP of $350 NmL_{CH4}/g_{COD}$ [45]. The S_I can then be calculated as the difference to 100 % multiplied by the initial COD concentration of the filtered sample.

The literature dealing with BMP, anaerobic biodegradability and digestibility is wide: a detailed review is provided in [46]. The principle of anaerobic respirometry for identification of COD fractions was developed as an analogy to aerobic respirometry used for the determination of influent COD fractions for activated sludge system models [47, 48]. It has already been successfully applied in several studies [16, 27, 29].

Blumensaat and Keller [49] was one of the first publications to implement the ADM1 for modeling of a two-stage AD process, and S_I measured in raw municipal sludge was estimated to be 1.5 % of total COD or 4.5 % of soluble COD. Huete et al. [19] suggested an ADM1-based methodology for the characterization of sewage sludge from a municipal wastewater treatment plant (WWTP) and found that S_I in the influent was about 12 % of soluble COD. In order to estimate these values, anaerobic biodegradability tests were performed with the filtered sewage sludge. Boubaker and Ridha [50] modeled the mesophilic anaerobic co-digestion of olive mill wastewater (OMW) with olive mill solid waste and quantified S_I to be 15 % of the soluble COD for OMW. Ekama [42] calculated S_I from a mass balance around a primary settling tank and found a value of 27 % of soluble COD for primary sludge.

Upflow anaerobic sludge bed (UASB) reactors have been developed for the treatment mainly of industrial wastewater with a distinct share of soluble COD [51]. Hinken et al. [52] modeled a UASB treating starch wastewater from an industrial wheat starch company, where soluble COD was more than three-quarters of total COD. Due to its excellent biodegradability, undegradable COD was determined to be about 0.7 % of total COD (= 0.9 % of soluble COD), but it was decreased to 0.5 % during model calibration. However, soluble COD-dominated applications are limited to industrial wastewater and their characteristics differ widely from industry to industry, for which reason comparison with municipal or other industrial wastewater is not possible.

The share of organic, but unbiodegradable compounds differs and is strongly dependent on the biomass input. Some aromatic [53, 54] and aliphatic compounds [55], among others, do not seem to be biodegradable under anaerobic conditions. Nevertheless, adaptation of microorganisms to degrade such compounds [56] as well as cometabolic degradation [57] should not be disregarded when comparing different values.

2.2 Determination of the Particulate Inert Fraction X_I

Similar to the soluble fraction, performing a BMP test with the particulate fraction of the input biomass would also be one possibility of determining X_I. In comparison to physical-chemical analysis, anaerobic respirometry allows insight into the rate-limiting step [16].

However, a physical-chemical analysis is applicable, because, in contrast to the soluble inert, the corresponding unbiodegradable compounds can be summarized as chemical parameters. Figure 4 depicts biomass characterization according to the extended Weender analysis [58] with van Soest fractions [59], which have been used in recent years as standard methods to characterize the chemical composition of energy crops, such as biodigester feedstock [60–62], as well as a simple and rapid method for ADM1 influent fractionation [31–33, 63–65].

Proteins, lipids, and starch are defined as totally digestible, but their degradation is regulated by the hydrolysis constant concerned. By contrast, lignin is not and

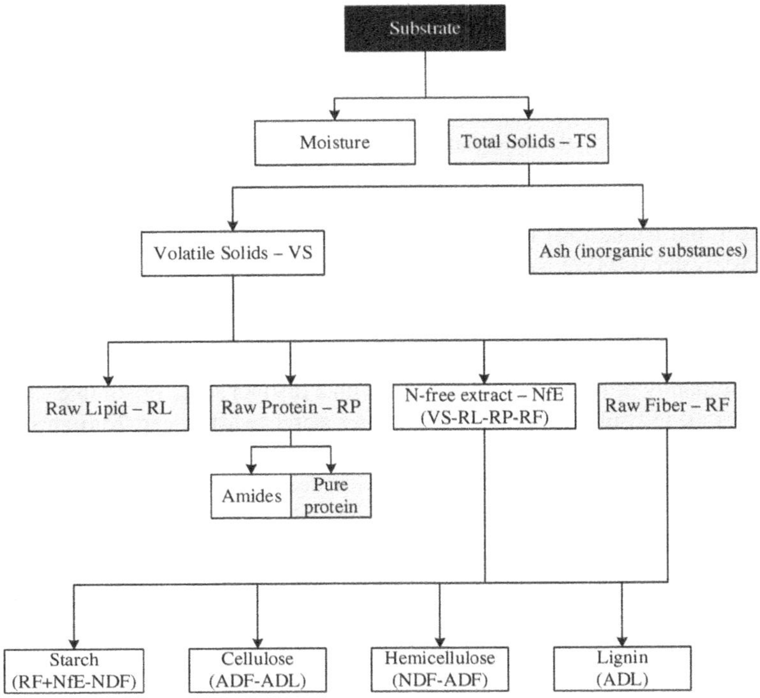

Fig. 4 Characterization of particulate input biomass by extended Weender analysis

hemicellulose and cellulose are only hardly anaerobically degradable. While all approaches agree that the particulate inert fraction X_I consists of lignin (ADL), handling of the moderate digestibility of cellulose (ADF − ADL) and hemicellulose (NDF − ADF) differs [31, 33]. One possibility is to introduce a slowly and readily degradable carbohydrate fraction with different hydrolysis rate constants, as has been proposed by several authors [66, 67]. Alternatively, the different isotopic composition, i.e. abundance of ^{13}C isotopes, of cellulose, lignin and protein in plant biomass [68] might allow to distinguish their specific degradability rates through the utilization of δ^{13} C measurements in the biogas [69]. Another possibility, and more pragmatic approach, is to divide the carbohydrate fraction into a fraction of starch (RF + NfE − NDF) and a degradable part of cellulose and hemicellulose ((NDF − ADL) ·d), while the rest is added to the inert fraction consisting of lignin (ADL) and the nondegradable part of cellulose and hemicellulose ((NDF − ADL) · (1 − d)). The factor d regulates the proportion and is generally calibrated with the gas curve. However, when the overall degradation level (D_{VS}, calculated, for instance, by the equation provided by [70]) is known, it can also be calculated assuming a total degradation of lipids (RL), proteins (RP), starch (RF + NfE − NDF) and degradable part of cellulose and hemicellulose ((NDF − ADL) · d) [33]:

$$\mathbf{d} = \frac{\mathbf{NDF} - \mathbf{VS} \cdot (1 - \mathbf{D_{VS}})}{\mathbf{NDF} - \mathbf{ADL}} \tag{1}$$

In contrast to S_I, which is usually in the lower percent range, typical values of X_I are presented in literature within a wide range between 3 and 60 % (see Table 7).

An alternative to the anaerobic respirometry method for S_I and the physical-chemical analysis for X_I is the calibration of the model upon the gas curve measured [71]. The method applied is rather simple and relies on the relation between biodegradable COD and methane production. Proving that all other influencing parameters, such as the disintegration or hydrolysis rate coefficients, are already defined, the share of unbiodegradable COD will be easily obtained by fitting the model to the gas flow measured. However, applying nonlinear parameter estimation techniques to both continuously operated full-scale and batch-operated lab-scale systems, the study in Batstone et al. [71] demonstrated that hydrolysis rate coefficients are not fully identifiable (unbound at higher values) from full-scale datasets. The full-scale system also indicated a higher degradability extent of the substrate treated than was estimated for the lab-system using the BMP test. The extent of degradability was found to be variable during the whole investigation study. This is of general importance when the degradability and, in turn, the unbiodegradable fraction, is assessed by BMP tests, when these are performed only once for a certain timestamp.

3 Determination of the Biodegradable Fraction

The composite particulate material in ADM1 is disintegrated with constant stoichiometric parameters into carbohydrates (30 %), proteins (30 %), lipids (30 %) and inerts (10 %).

Carbohydrates Carbohydrates are an important class of macromolecules with cellulose as probably the most common member of this fraction, since it is, firstly, an important structural component of the primary cell wall of plant materials, and, secondly, makes up about 30–50 % of suspended solids in wastewater (Fig. 5).

Lipids, i.e. phospholipids (Fig. 6), are amphipathic macromolecules (hydrophilic and hydrophobic) and essential components in the cells of the *Bacteria*, *Eukarya*

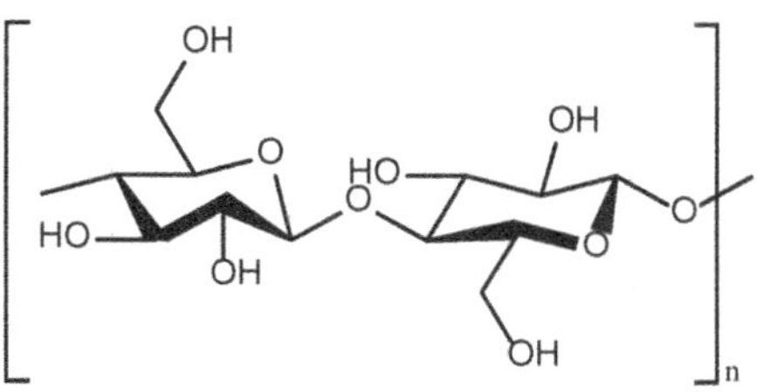

Fig. 5 Chemical structure of a cellulose molecule, $C_6H_{10}O_5$

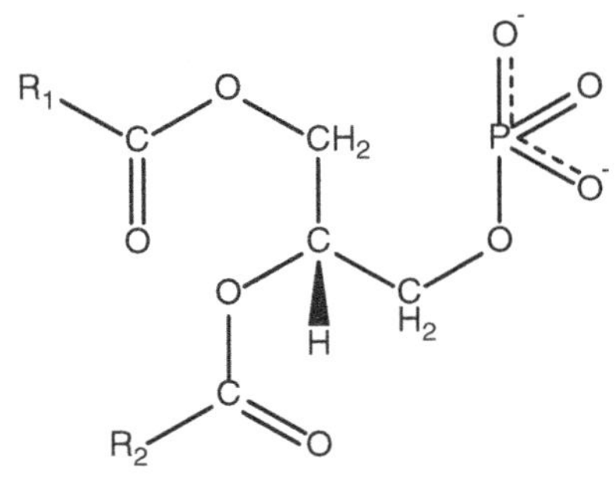

Fig. 6 Chemical structure of phosphatidic acid $C_7H_{11}PO_8^-$—the simplest phospholipid

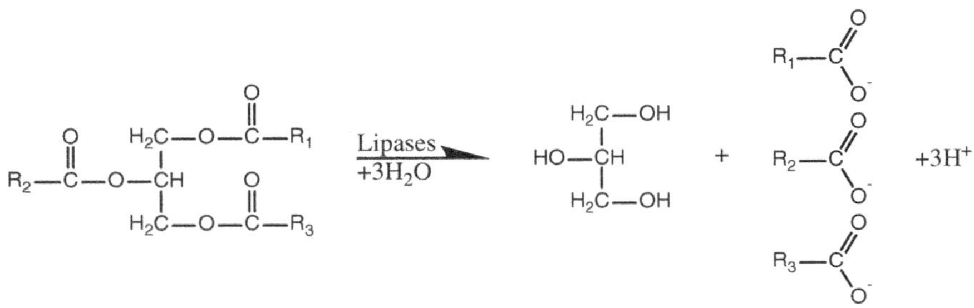

Fig. 7 Hydrolysis of triacylglycerol into glycerol and long chain fatty acids [73]

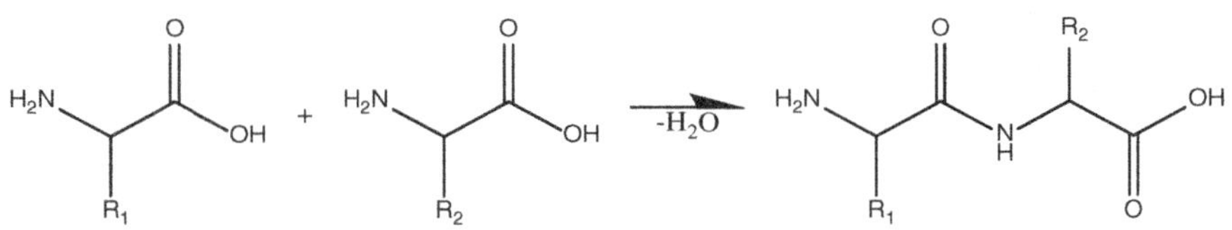

Fig. 8 Dipeptide formed out of two AA with elimination of a water molecule. This simple reaction between AA forms the basis for a fictive chemical formula for proteins that is $C_6H_{12}O_3N_2$

and *Archaea*. They are composed of a polar compound, such as glycerol, that is bonded to fatty acids or by other hydrophobic molecules by ester or ether linkage [72].

Lipids in wastewater are usually represented in the form of triacylglycerides and underlie hydrolysis through exo-enzymes, which are produced by the acidogenic bacteria. A schematic presentation of the hydrolysis step for lipids is shown in Fig. 7.

Proteins are the most abundant class of macromolecules in cells. They are about 50 % of the dry weight of most cells and made up of amino acids (AA) from ribosomes. They form polymers covalently bonded by peptide bonds with a dipeptide as the simplest form and water as a byproduct (Fig. 8). Proteins can

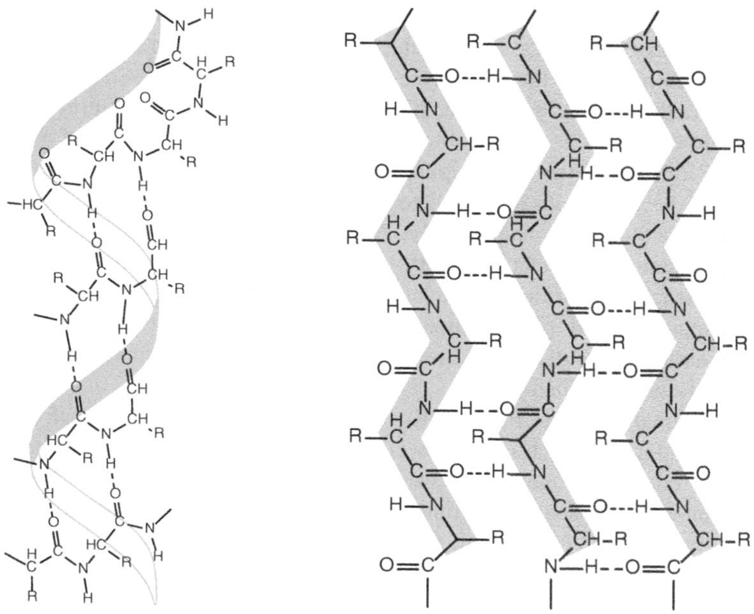

Fig. 9 Secondary structure of polypeptides [72]. *Left* α-helix secondary structure. *Right* β-sheet secondary structure. Hydrogen bonding occurs between electronegative oxygen atoms and hydrogen atoms

contain as few as a minimum of ten AA that constitute a polypeptide chain or as many as thousands, forming the so-called primary structure [74]. Consequently, enormous variation in protein structure and, thus, function is possible [72].

When a polypeptide is formed, it does not remain in its primary linear structure, but folds instead, forming a more stable complex known as the secondary structure (Fig. 9). Thus, one distinguishes between the geometrical arrangements of helices (α structures), strands (β sheets) and coils [75]. The chemical bonds that play an important part here are weak noncovalent linkages known as hydrogen bonds. Finally, the whole three-dimensional complex and irregular folding of any peptide chain is given by its tertiary structure, while the so called quaternary structure describes the interplay between various peptide chains that make up a whole protein [75].

Considering the variations of the individual fractions on the chemical level, it becomes clear why a more precise definition of composite particulate material in ADM1 is meaningful and that the lumped definition of the single compounds, carbohydrates, lipids and proteins, will not reflect its complexity.

Most of the published approaches performing a detailed influent fractionation aim at characterizing the influent composition by assuming a correlation between analytical measurements and the substrate composition and fractions. Total and soluble proteins, for example, can be roughly determined according to the Kjeldahl method or, in more detail, by the Lowry method [76]. Another, and perhaps the most detailed procedure, is to analyze the total AA composition of a substrate, from which the sum results in the total proteins. Lipids can be analyzed by the Soxhlet extraction method [77, 78], which is the method found in literature most used. Other methods are still conceivable, e.g. Roese-Gottlieb, Bligh and Dyer, and Modified Bligh and Dyer extraction methods [79]. The total and soluble carbohydrates can be measured according to the Bertrand method and to the Anthrone colorimetric method [80], respectively. An extended Weender analysis, as described in Naumann and Bassler [58] and van Soest [59], is commonly performed for fibrous feedstock. In addition to the latter, and as cellulose plays an important role in the fermentation of plant biomass, specific methods are available, e.g. the Updegraff assay [81] and, more recently, a rapid determination method for cellulose is given in Bauer and Ibáñez [82].

The analytical methods listed will result in a concentration unit of the corresponding compound, which is either weight per volume (w/v) or weight per weight (w/w). The challenge one meets using physical-chemical analysis for influent fractionation is to subsequently transfer the analytical results to the model units, which is, for the organic material, the COD.

3.1 Transfer of Physical-Chemical Analyses to the Basic Model Unit

Eckenfelder and Weston [83] postulated and proved proportionality between COD and VSS experimentally. Servizi and Bogan [84] described proportionality between the oxygen demand of microorganisms and the released energy, expressed as COD. Hereby, the basis was found for the establishment of COD as a measuring unit for mathematical models describing biological wastewater treatment processes. The COD has two advantages, as it allows, on the one hand, an indirect assessment of the magnitude of the organic pollution and, on the other hand, a closed mass balance for the interconnected transformation processes. All organic ADM1 model components are referred to a COD basis. The COD concept was criticized in Kleerebezem and van Loosdrecht [85], and using a mole-based description for the fermentation processes was recommended. An ADM1 implementation in molar units has been provided in Rodríguez et al. [86], which has been enhanced in Penumathsa et al. [87] and applied to modeling biohydrogen production. One reason for choosing a COD-based model description in ADM1 was that this concept allows the model to be connectable to other modeling frameworks, such as the activated sludge model (ASM) series [25]. While COD is an established parameter used for wastewater characterization, organic matter from agricultural origin or municipal solid wastes are almost exclusively determined by the VS content. Fibrous material in particular is poorly accessible to COD analysis. In order to bypass this issue, some studies dealing with the mono-fermentation of energy crops and liquid manure transferred, in accordance with Eckenfelder and Weston [83], the

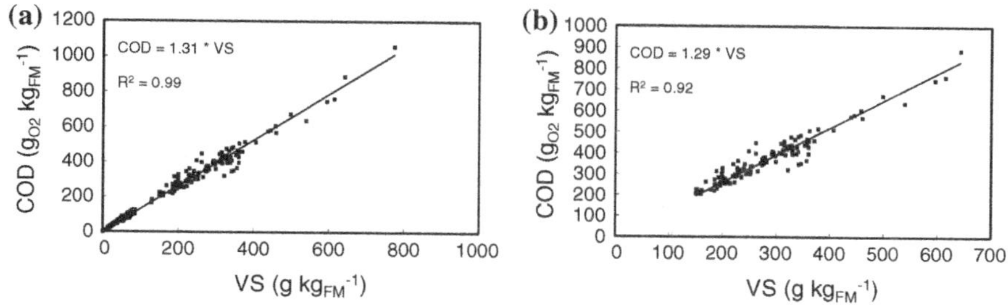

Fig. 10 Correlation between COD and VS for **a** liquid manure [31] and **b** grass silage [32]

analytically measured VS to COD units by establishing an empirical correlation between both parameters (Fig. 10).

The correlation coefficient derived for liquid manure has a similar magnitude to grass silage. Grass silage contains only a limited part of readily degradable carbohydrates, such as starch. The readily degradable fraction of carbohydrates in liquid manure is similarly low, as the rumen of the cow can be seen as a pre-fermenter. However, the conclusion that both sorts of biomass possess the same composition of cellulose, proteins and fats is erroneous. It is rather that a different combination of proteins and lipids leads to a similar correlation coefficient. The establishment of empirical correlations between basic analytical parameters is an easy method to achieve COD units, but has to be performed for each substrate mix separately. It has significant advantages for inhomogeneous substrate mixes, for example, combined fermentation of liquid manure and different cosubstrates, where the typical elemental composition is unknown and can vary strongly depending on the use of the cosubstrate.

An alternative method is to calculate the ThOD at a known or, respectively, chosen elemental composition $C_aH_bO_cN_d$ according to the formula:

$$\mathbf{ThOD} = \frac{16 \cdot (2\mathbf{a} + 0.5(\mathbf{b} - 3\mathbf{d}) - \mathbf{c})}{12\mathbf{a} + \mathbf{b} + 16\mathbf{c} + 14\mathbf{d}} \left[\frac{g_{O_2}}{g_{C_aH_bO_cN_d}}\right] \tag{2}$$

The elemental composition of carbohydrates is generally chosen to be $(C_6H_{10}O_5)_n$ [19, 27, 33, 44]. The typical elemental composition of lipids is referred to $C_{51}H_{98}O_6$ [17, 19], $C_7H_{11}PO_8^-$ [27] or $C_{57}H_{104}O_6$ [33, 44]. The highest variety in selecting a typical elemental composition exists for the proteins, $C_1H_{2.52}O_{0.87}N_{0.26}$ [17], $(C_4H_{6.1}O_{1.2}N)_n$ [19], $C_6H_{12}O_3N_2$ [27] or $C_5H_7O_2N$ [33, 44]. It is also conceivable to refer to a protein which is predominant within a substrate. Zein, for instance, is a class of prolamine protein and is the major protein of maize (*Zea mays*) [88]. The chemical composition of zein can easily be calculated, based on the elemental analysis given in Dennstedt and Haßler [89], to $C_{317.02}H_{505.90}O_{95.27}N_{80.12}S_{1.68}$. Hence, the ThOD of proteins calculated from Eq. (2) depends on the basic formula chosen. However, any chemical formula chosen will never truly reflect the complex composition of proteins. These are composed of 21 different AA, and only the fully determination of this pool allows a realistic calculation of the organic acids stoichiometrically produced via protein/AA degradation.

3.2 Determination of Amino Acids Composition

Ramsay and Pullammanappallil [90] introduced a general approach of determining the stoichiometric coefficients for protein degradation using the example of casein. Their algorithm is based upon Stickland reactions and the AA composition of casein. The method provided is explained in Table 1 for maize silage (shown on our own data as an example).

Though Ramsay and Pullammanappallil [90] used literature values for the AA content of casein, it would be more advisable to determine protein AA individually for the substrate applied via chemical analysis. The results should be calculated in terms of mole AA per C-mole of substrate. From here on, the molecular formula for casein ($CN_{0.23}H_{1.9}O_{0.51}$) was compiled. For certain substrates it is also possible to take up whole AA composition charts from literature that result in an overview of different protein compositions for a diverse group of substrates (Table 2).

Table 3 clearly shows that the stoichiometric coefficients, calculated for acetic acid to valeric acid, differ in parts significantly from those provided by the original ADM1 publication as default values for the degradation of AA. A detailed characterization of the AA state variable (lumped as S_{aa}) is inevitable, especially when ADM1 is used to estimate valeric acid production. Valeric acid is only produced during AA fermentation in ADM1. Fermentation of protein rich substrates often leads to instabilities in digester operation, leading to an increase in VFA concentration [56, 96–98]. The estimation of organic acid production will be improved, as it is not only a matter of parameter calibration, when a more detailed characterization for AA is performed. However, the literature is undertheorized within this context. The ADM1 state variable S_{aa} can still be regarded as a lumped parameter, comprising the whole pool of single AA. Influent characterization can be indirectly performed via the calculation of the stoichiometric coefficients for single organic acids (C_2–C_5), as demonstrated in Table 3 using literature data.

4 Quantification of Active Biomass Within the Influent

The characterization and quantification of the active biomass in the reactor is less important for continuously operated systems, since the initial biomass concentration can be calibrated from the measurements in steady state. A commonly used approach is to perform a simulation prior to the real one, with default values for the initial biomass concentrations. This pre-simulation is run with the same influent, but is only used to generate the biomass concentrations in steady state, which can then be applied as the initial biomass concentrations for the real simulation run.

Table 1 Calculation of stoichiometric coefficients for protein degradation of maize silage according to Ramsay and Pullammanappallil [90], using our own data for AA composition of maize as an example

Amino acid (AA)	Content (mole AA/c-mole protein)	C_2 acid (mole/mole AA)	C_3 acid (mole/mole AA)	C_4 acid (mole/mole AA)	C_5 acid (mole/mole AA)
Alanine	0.0311	1.0			
Arginine	0.0036	0.5	0.5		0.5
Aspartic acid	0.0147	1.0			
Cysteine	0.0034	1.0			
Glutamic acid	0.0313	1.0		0.5	
Glycine	0.0183	1.0			
Histidine	0.0079	1.0		0.5	
Isoleucine	0.0109				1.0
Leucine	0.0221				1.0
Lysine	0.0060	1.0		1.0	
Methionine	0.0017		1.0		
Phenylalanine	0.0081				
Proline	0.0160	0.5	0.5		0.5
Serine	0.0099	1.0			
Threonine	0.0086	1.0		0.5	
Tryptophan	0.0000				
Tyrosine	0.0056	1.0			
Valine	0.0131			1.0	
TOTAL (α)	Mole/C-mole	0.1466	0.0155	0.0430	0.0427

Two acid hydrolyses were carried out for each sample of maize (*zea mays*): one with preceding oxidation and one without oxidation. The preceding oxidation step was conducted to detect cysteine as cysteine acid and to oxidize methionine to methsulfone. After oxidation and overall hydrolyses, each sample was analyzed by AA analysis (ASA) via pre-column derivatization and HPLC. The ASA was carried out with a polymer cation-exchanger column with a particle size of 4 µm (125 x 4 mm ID) by using the AA analyzer LC3000 (Eppendorf-Biotronic, Hamburg, Germany), post-column derivatization with ninhydrin at 125 °C and a final photometric detection at 570 nm

Assessing the initial biomass concentration for batch tests is much more important, because, in this case, no steady state will be reached due to the lack of biomass added during the fermentation process. One possibility to address the problem is to take the initial biomass concentrations for the batch experiments from the results of the steady state simulation of the continuous system, from which the inocula were sampled [16, 99].

Alternatively, the initial biomass concentration can be roughly estimated by molecular biological tools, such as fluorescence in situ hybridization (FISH) or polymerase chain reaction denaturing gradient gel electrophoresis (PCR-DGGE). The former has been applied in Lübken et al. [31] for the simulation of a manure-based co-digestion system. However, due to the very small volume of the

Table 2 Chemical compositions of proteins calculated for various substrates based upon AA charts given in literature

Substrate	Chemical composition of protein	References
Galantine	$CH_{2.0461}O_{0.5719}N_{0.3170}S_{0.0022}$	[91]
Maize silage[a]	$CH_{1.9528}O_{0.5408}N_{0.2448}S_{0.0051}$	Own data
Grass silage[a]	$CH_{1.9821}O_{0.5470}N_{0.2507}S_{0.0028}$	Own data
Rye	$CH_{1.9376}O_{0.5616}N_{0.2513}S_{0.0093}$	[92]
Chlorella vulgaris (green alga)	$CH_{1.9230}O_{0.4966}N_{0.2559}S_{0.0072}$	[93]
Scenedesmus obliquus (green alga)	$CH_{1.9565}O_{0.5147}N_{0.2703}S_{0.0047}$	[93]
Blue algae	$CH_{1.9947}O_{0.5557}N_{0.2558}S_{0.0033}$	[94]
Casein	$CH_{1.9028}O_{0.5138}N_{0.2302}S_{0.0050}$	[90]
Egg	$CH_{1.8903}O_{0.4999}N_{0.2427}S_{0.0106}$	[93]
Soybean	$CH_{1.9152}O_{0.4896}N_{0.2571}S_{0.0065}$	[93]
Sweet potato	$CH_{1.8820}O_{0.5364}N_{0.2423}S_{0.0095}$	[95]

Calculations were carried out according to the algorithm provided in Ramsay and Pullammanappallil [90]. The associated stoichiometric coefficients (α) are given in Table 3
[a]Based on our own analytical determination of AA composition

Table 3 Overview of theoretical stoichiometric coefficients (α) for selected protein mixtures

Fermentation product	Acetic acid ($f_{ac,aa}$)	Propionic acid ($f_{pro,aa}$)	Butyric acid ($f_{bu,aa}$)	Valeric acid ($f_{va,aa}$)
ADM1[a]	0.400	0.050	0.260	0.230
Blue algae	0.325	0.041	0.242	0.293
Casein	0.279	0.072	0.280	0.299
Chlorella vulgaris (green alga)	0.301	0.064	0.271	0.278
Egg	0.273	0.068	0.260	0.309
Galantine	0.399	0.120	0.186	0.293
Maize silage	0.321	0.044	0.236	0.304
Grass silage	0.334	0.045	0.236	0.292
Rye	0.328	0.078	0.264	0.272
Scenedesmus obliquus (green alga)	0.315	0.063	0.265	0.275
Soybean	0.287	0.063	0.290	0.298
Sweet potato	0.319	0.055	0.257	0.262

[a]Suggested as default values

sample processed and the questionable representativeness in the fixing, staining and counting of the cells, the method is rather semi-quantitative. Recent advances in the use of molecular biology tools have greatly improved the understanding of the AD process and the function of the microorganisms involved [100]. Due to the recent progress of pyrosequencing and other next-generation sequencing techniques [101], the analysis of a considerably increased number of sequences is becoming affordable and will probably help to better characterize the initial biomass concentration in the future.

The transfer of bacterial biomass concentration, if known, to COD units could be easily performed using the ThOD applied to the widely accepted elemental composition of biomass, which is $C_5H_7NO_2$ [102] (113 $g_{biomass}$ per mole). Some literature resources refer to different elemental compositions, e.g. $C_5H_9NO_3$ [103] or $CH_{1.8}O_{0.5}N_{0.2}$ [104].

5 Calculation Frameworks for the Derivation of ADM1 Input State Variables

The literature overview reveals the fact that ADM1 shows weaknesses related to the substrate characterization, and this has driven the development of model add-ons and modifications forward. Four selected approaches for this purpose will be presented in the following paragraphs, as listed in Table 4. The approaches were classified according to (i) extended Weender analysis, (ii) elemental analysis, (iii) elemental analysis for high solids waste and (iv) anaerobic respirometry. The coupling of the ADM1 input state variables with ASM-type models is not referred to as these approaches focus more on enabling plant-wide modeling concepts for WWTPs [25, 28, 105].

Extended Weender analysis The substrate characterization in Lübken et al. [31] proposes that the COD should be replaced by VS when analyzing manure-based co-digestion systems. In this particular study, substrate characterization was based upon practical measurements, similarly upon methods according to van Soest and Wine [59] and Naumann and Bassler (Weender analysis) [58], to characterize the substrate in terms of carbohydrates, proteins and fats. This method assumes that a portion of carbohydrates (starch, cellulose, hemicellulose and lignin) is inert to AD, while proteins and lipids are fully biodegradable. The determination of the ADM1 inflow contents of proteins, lipids, carbohydrates and inerts is, therefore, quite straightforward using Eq. 3–6.

Table 4 Calculation frameworks. Explanations of abbreviations are given in the following paragraphs

Analytical measurements	References
TS, VS, COD, VFA, pH, NH_4-N, RP, RF, RL, NfE, NDF, ADF, ADL	[31]
COD, TOC, N_{org}, Alkalinity	[17]
COD_p, COD_s, VFA, TOC, TIC, N_{org}, TAN, TP, ortoP, S_{cat}, FS	[27]
TS, VS, TKN, TAN, COD	[16]

RP, RF, RL, NfE, NDF, ADF and ADL were determined through extended Weender analysis

$$\mathbf{X_{pr}}\left[\frac{\mathbf{kg_{COD}}}{\mathbf{d}}\right] = \mathbf{Q_{in}} \cdot \mathbf{RP} \cdot \mathbf{TS} \cdot \mathbf{i_{COD/VS}} \tag{3}$$

$$\mathbf{X_{li}}\left[\frac{\mathbf{kg_{COD}}}{\mathbf{d}}\right] = \mathbf{Q_{in}} \cdot \mathbf{RL} \cdot \mathbf{TS} \cdot \mathbf{i_{COD/VS}} \tag{4}$$

$$\mathbf{X_{ch}}\left[\frac{\mathbf{kg_{COD}}}{\mathbf{d}}\right] = \mathbf{Q_{in}} \cdot \left[\mathbf{RF} + \mathbf{Nfe} - \mathbf{ADL} - (\mathbf{ADF} - \mathbf{ADL})_{\mathbf{not\ degradable}}\right] \cdot \mathbf{TS} \cdot \mathbf{i_{COD/VS}} \tag{5}$$

$$\mathbf{X_{i}}\left[\frac{\mathbf{kg_{COD}}}{\mathbf{d}}\right] = \mathbf{Q_{in}}\left[\mathbf{ADL} + (\mathbf{ADF} - \mathbf{ADL})_{\mathbf{not\ degradable}}\right] \cdot \mathbf{TS} \cdot \mathbf{i_{COD/VS}} \tag{6}$$

The coherences within these equations can easily be accessed by means of Fig. 4, which partitions the substrate analysis basically into six important analyzable categories (see Fig. 4, grey boxes): TS, ash, RP, RL, raw fiber (RF) and nitrogen-free extracts (NfE).

In the course of substrate analysis, water and dry matter in terms of TS are distinguished. The TS are expressed on a mass percent (%) basis and further split up into VS and ash. The analytical determination of TS and VS is straightforward according to standard methods. As the dry content of the substrate is, by definition, TS, VS are reported as a percentage of TS (% TS). The residual fraction consists of inorganic substances that are summarized under the term "ash." This fraction contains macro and trace elements, sand, clay and pure ash. An extended Weender analysis is performed to measure organic matter in a more detailed way in terms of RL, RP, RF and NfE. All those parameters are expressed as percentage of TS (% TS). Raw proteins can be further distinguished into amides, for instance, free AA, acid amides or peptides. The residue is, consequently, pure protein. The term "NfE" is a bit misleading, as it has nothing to do with nitrogen, nor it is an extract. The term refers to the soluble carbohydrate of the feed and is the only component which is not determined analytically, but is calculated by difference. Carbohydrates are further subdivided into starch (RF + NfE − NDF), cellulose (ADF − ADL) and lignin (ADL). Hemicellulose is approximately analyzed by van Soest and Wine fractions as the difference between neutral detergent fiber (NDF) and acid detergent fiber (ADF).

As is always the case, each analytical step has smaller errors or bottlenecks that shall be briefly discussed. A smaller source of errors is seen in the determination of ash and raw lipids. For ash, errors may arise when temperatures are too high, leading to a volatilization of elements, such as chloride, zinc, selenium or iodine. In this case, the ash determination tends to underestimate mineral contents. The determination (or extraction) of raw lipids is mostly performed using an anhydrous diethyl ether. In this case, it is assumed that all substances that are soluble in ether are fats. This is not always the case, as, for instance, plant pigments are also soluble in ether, but do not possess the same nutritional value of fats. Nonetheless, the errors for ash and lipid determination obtained are relatively small.

The determination of RF possesses the largest source of errors. At first, the application of acid and base (mostly sulfuric acid and potassium hydroxide) solubilize some of the true fiber, in particular, hemicellulose, pectin and lignin. Secondly, cellulose is also partially lost, hence, RF underestimates the true fiber content.

Another bottleneck is seen in the determination of NfE that accumulates all of the errors that exist in other proximate analysis, as it is determined by subtracting raw lipids, RP and RF from VS. In the case of incorrect analysis of the previous components, mistakes will proceed up to this calculation step. This refers in particular to the determination of RP. In this context, it is assumed that all nitrogen is present in the sample in protein form, which is not always necessarily true. Nitrogen could be present in the form of nucleic acids (DNA, RNA) or urea (CH_4N_2O). Moreover, proteins require different correction factors, since they have different AA sequences. Raw protein is mostly determined by the Kjeldahl method.

The main advantage of this calculation framework is that the extended Weender analysis is nowadays often routinely performed in the agricultural sector for animal husbandry and has also been established on agricultural biogas plants. The demands of ADM1 in respect of substrate characterization is generally regarded as hardly to be fulfilled by routine analysis. With the extended Weender analysis, a routine analysis for agricultural feedstock exists which provides detailed information about single substrate constituents. However, the influence of analytical errors of the multistep analysis procedure has to be considered when model parameter estimation is performed. Uncertainties in the influent fraction are generally neglected in literature dealing with parameter estimation and uncertainty analysis. The basic methodology, as described above, has been adapted to other lignocellulosic biomass, such as grass silage [32, 33], as this approach enables a quite flexible handling.

Elemental analysis Another calculation framework for substrate characterization is the one introduced in Kleerebezem and Van Loosdrecht [17]. The authors developed an algorithm to calculate the lumped elemental composition of the organic substrates in wastewater from a limited number of widely available analyses, such as COD, TOC, N_{org}, Alk_{VFA} and Alk_{IC} (Fig. 11).

The results of the practical chemical measurements flow into the calculation of the elemental composition of the organic substrate (Eq. 7) using Eqs. (8–11).

$$\mathbf{C_xH_yO_zN^{-III}{}_v^u} \tag{7}$$

$$\mathbf{y} = \frac{2 \cdot \mathbf{COD} + \mathbf{Alk_{VFA}} - 2 \cdot \mathbf{N_{org}}}{\mathbf{TOC}} \tag{8}$$

$$\mathbf{z} = 2 - \frac{\mathbf{COD} + 0.5 \cdot \mathbf{N_{org}}}{\mathbf{TOC}} \tag{9}$$

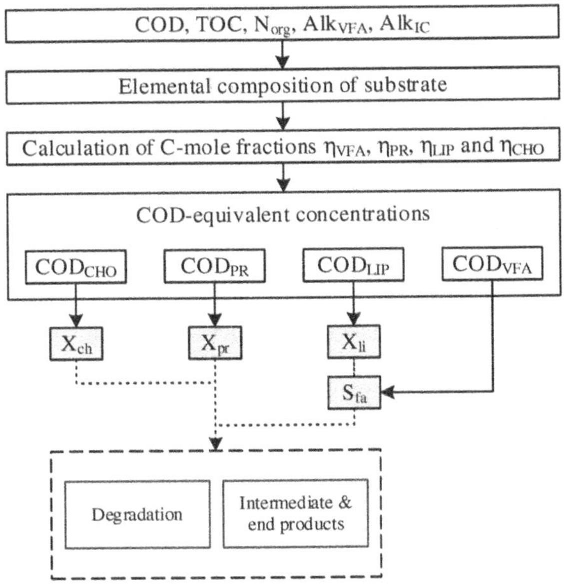

Fig. 11 Algorithm of waste characterization for implementation in ADM1 according to Kleerebezem and Van Loosdrecht [17]

$$\mathbf{v} = \frac{\mathbf{N_{org}}}{\mathbf{TOC}} \tag{10}$$

$$\mathbf{u} = \frac{-\mathbf{Alk_{VFA}}}{\mathbf{TOC}} \tag{11}$$

From here on, the carbon mole fractions for proteins, carbohydrates, lipids and VFAs are calculated using the elemental substrate composition and Eqs. (12–15).

$$\eta_{\mathbf{VFA}} = \frac{\mathbf{u}}{\mathbf{Ch_{VFA}}} \tag{12}$$

$$\eta_{\mathbf{PR}} = \frac{\mathbf{v}}{\mathbf{N_{PR}}} \tag{13}$$

$$\eta_{\mathbf{LIP}} = \frac{\mathbf{y} - 2 \cdot \mathbf{z} - 3 \cdot \mathbf{v} - \mathbf{u}}{4 \cdot \left(\gamma_{\mathbf{LIP}}^{-1}\right)} \tag{14}$$

$$\eta_{\mathbf{CHO}} = 1 - \eta_{\mathbf{LIP}} - \eta_{\mathbf{VFA}} - \eta_{\mathbf{PR}} \tag{15}$$

As mentioned previously, ADM1 requires COD-based substrate concentrations, thus, the carbon mole fractions calculated have to be converted into COD equivalents using Eqs. (16–19).

$$\mathbf{COD_{CHO}} \cdot \left[\frac{\mathbf{g_{COD}}}{\mathbf{L}}\right] = \mathbf{TOC} \cdot \eta_{\mathbf{CHO}} \frac{\gamma_{\mathbf{CHO}}}{4} \cdot \mathbf{MW_{O_2}} \tag{16}$$

$$\mathbf{COD_{PR}} \left[\frac{\mathbf{g_{COD}}}{\mathbf{L}}\right] = \mathbf{TOC} \cdot \eta_{\mathbf{PR}} \cdot \frac{\gamma_{\mathbf{PR}}}{4} \cdot \mathbf{MW_{O_2}} \tag{17}$$

$$\mathbf{COD_{LIP}} \left[\frac{\mathbf{g_{COD}}}{\mathbf{L}}\right] = \mathbf{TOC} \cdot \eta_{\mathbf{LIP}} \cdot \frac{\gamma_{\mathbf{LIP}}}{4} \cdot \mathbf{MW_{O_2}} \tag{18}$$

$$\mathbf{COD_{VFA}} \left[\frac{\mathbf{g_{COD}}}{\mathbf{L}}\right] = \mathbf{TOC} \cdot \eta_{\mathbf{VFA}} \cdot \frac{\gamma_{\mathbf{VFA}}}{4} \cdot \mathbf{MW_{O_2}} \tag{19}$$

Nonetheless, the method's main limitation lies in averaging the practical measurements that will limit an ADM1 application to one single composite particulate (X_c) with constant composition, thus, dynamic changes of the waste characteristics will not be considered [27]. Moreover, the authors based their waste characterization algorithm upon a protein mixture from slaughterhouse waste. However, as the protein composition differs from one substrate to another, it would be advisable to determine the protein mixture with much more precision (see Chap. 3.2).

Elemental analysis for high solids waste Based upon the algorithm mentioned previously and other studies, Zaher et al. [27] developed a procedure for substrate characterization that considers a more extended list of practical measurements: total COD (COD_t), soluble COD (COD_s), VFA, total carbon (TC), total inorganic carbon (TIC), total Kjeldhal nitrogen (TKN), total ammonia-nitrogen (TAN), total phosphorous (TP), orthophosphate (orthoP), total alkalinity (S_{cat}), TS and total VS (TVS). The procedure was originally developed to construct interfacing models between subsystems considered in wastewater treatment and uses a continuity-based interfacing method (CBIM) as the basis and a Petersen presentation to describe the transformation matrix between [105]. Due to the CBIM, it is also possible to yield the state variables of ADM1 and, thus, their specific elemental composition.

The authors used the general CBIM, which was updated to interface the ADM1 with practical measurements adopted and slightly extended from Kleerebezem and Van Loosdrecht [17]. The final transformation matrix was designed to estimate the substrate composition in ten conversions maintaining the continuity of the ThOD, all elements and charge intensity [27]. The conversion steps comprise (1) ammonia, (2) bicarbonate, (3) orthoP, (4) cations, (5) VFAs, (6) sugars, (7) lipids, (8) proteins, (9) carbohydrates and (10) organic inerts. All of these classes are assigned a parameter from practical measurements where a dependency is considered.

Nonetheless, not every parameter from the set of practical measurements can be used straightforwardly. Hence, one has to start with a rearrangement to represent unique components for which the elemental mass fractions are assumed. The components in the final matrix comprise particulate COD (COD_p), COD_s − VFA, VFA, TOC, N_{org}, TAN, TP-orthoP, orthoP, TIC, S_{cat} and FS. The COD_p is calculated as the difference between COD_t and COD_s. The TOC is calculated as the difference between TC and TIC. Organic nitrogen and TP are calculated similarly

Table 5 Matrix description of the interface between the Peterson matrix of practical measurements and ADM1 [105]

Peterson matrix of practical measurements					Petersen matrix of ADM1				
	X_1	X_2	…	X_P	X_{P+1}	X_{P+2}	…	X_{P+Q}	Rate
Conv. 1	$\nu_{1,1}$	$\nu_{1,2}$	…	$\nu_{1,P}$	$\nu_{1,P+1}$	$\nu_{1,P+2}$	…	$\nu_{1,P+Q}$	ρ_1
…	…	…	…	…	…	…	…	…	…
Conv. n	$\nu_{n,1}$	$\nu_{n,2}$	…	$\nu_{n,P}$	$\nu_{n,P+1}$	$\nu_{n,P+2}$	…	$\nu_{n,P+Q}$	ρ_n
Composition matrix of practical measurements					Composition matrix of ADM1				
ThOD	$i_{ThOD,1}$	$i_{ThOD,2}$	…	$i_{ThOD,P}$	$i_{ThOD,P+1}$	$i_{ThOD,P+2}$	…	$i_{ThOD,P+Q}$	
C	$i_{C,1}$	$i_{C,2}$	…	$i_{C,P}$	$i_{C,P+1}$	$i_{C,P+2}$	…	$i_{C,P+Q}$	
N	$i_{N,1}$	$i_{N,2}$	…	$i_{N,P}$	$i_{N,P+1}$	$i_{N,P+2}$	…	$i_{N,P+Q}$	
H	…	…	…	…	…	…	…	…	
O	…	…	…	…	…	…	…	…	
P	…	…	…	…	…	…	…	…	
Charge	$i_{e.,1}$	$i_{e.,2}$	…	$i_{e.,P}$	$i_{e.,P+1}$	$i_{e.,P+2}$	…	$i_{e.,P+Q}$	
Covalent bond	$i_{b,1}$	$i_{b,2}$	…	$i_{b,P}$					

Each conversion X_P in the table depicted is characterized by its "stoichiometry" $\nu_{n,P}$

from the difference between TKN and TAN and from the difference between TP and orthoP. The VFAs, TAN, orthoP, TIC and S_{cat} derive straightforwardly from the practical measurements.

With these assumptions, the matrix can now be composed. The upper left pane of the composition matrix (Table 5) is calculated by taking the COD equivalent multiplied by the number of carbon atoms and the molecular weight of carbon. In the case where a component is used up, it will be given a negative sign (-). As an example, one mole of acetate equals 64 g_{COD}, owns two carbon atoms, while the molecular weight of carbon equals 12.01 g • mol^{-1}. Thus, −0.375 $g_C \cdot m^{-3}$ is entered into the composition matrix for TOC/VFA. The lower left pane lists the mass of elemental composition per stoichiometric unit of each component. Staying with the example of acetate, one mole is equivalent to 64 g_{COD}, having two oxygen atoms, thus, 0.5 $g_O \cdot g_{COD}^{-1}$ acetate is achieved. In a similar manner one calculates 0.0469 $g_H \cdot g_{COD}^{-1}$. The lower right pane derives from the original ADM1 composition matrix.

The actual application of the transformation matrix requires only the upper pane and is based on balancing the input from the practical measurements. Table 6 gives an excerpt of the final transformation matrix calculated by Zaher et al. [27] (please refer to the original paper for the full table). Taking the conversion of lipids as an example, 1 g • m^{-3} of organic phosphorus (TP-orthoP) is equivalent to 0.006458 $kg_{COD} \cdot m^{-3}$ of lipids (X_{li}) in ADM1, based upon the phosphorus balance. This also works in reverse order: 0.006458 $kg_{COD} \cdot m^{-3}$ of lipids (X_{li}) is equivalent to a TOC value of 2.71 g_C • m^{-3} and to a COD_p of 6.458 g_{COD} • m^{-3}.

Table 6 Excerpt from the calculated transformation matrix of the ADM1 interface to practical measurements, according to Zaher et al. [27]

		Practical measurements							ADM1		
		COD_p	DOD_s-VFA	VFA	TOC	N_{org}	TAN	TP-orthoP	X_{ch}	X_{pr}	X_{li}
		$g_{COD}\,m^{-3}$	$g_{COD}\,m^{-3}$	$g_{COD}\,m^{-3}$	$g_C\,m^{-3}$	$g_N\,m^{-3}$	$g_N\,m^{-3}$	$g_P\,m^{-3}$	$kg_{COD}\,m^{-3}$	$kg_{COD}\,m^{-3}$	$kg_{COD}\,m^{-3}$
1	Ammonia						−1				
2	Bicarbonate										
3	orthoP										
4	Cations										
5	VFA			−1	−0.375001						
6	Sugars		−1		−0.357						
7	Lipids	−6.457862			−2.712302			−1			0.0064579
8	Proteins	−6.857143			−2.57143	−1				0.0068571	
9	Carbohydrates	−1			−0.375				0.001		
10	Organic inerts	−1			−0.408601	−0.058001		−0.006448			

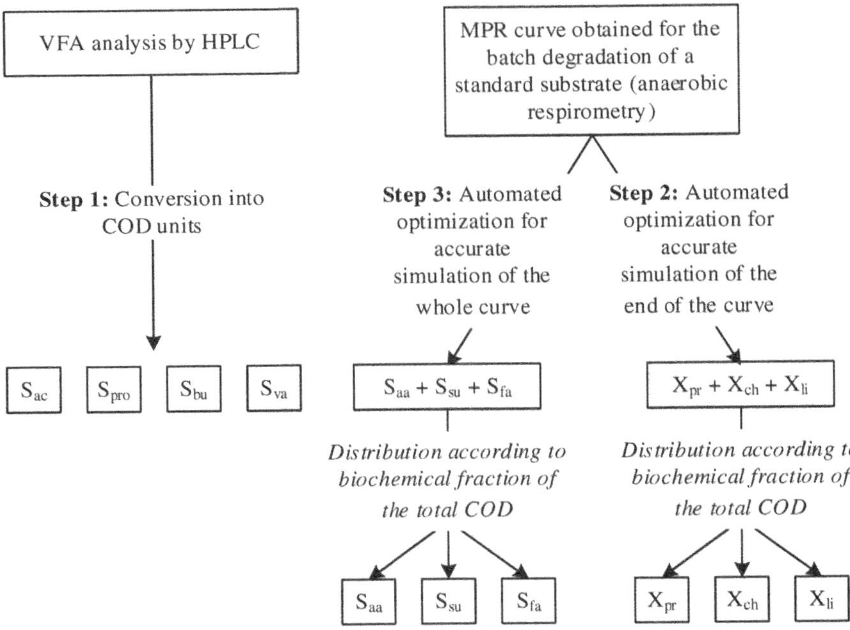

Fig. 12 Scheme for the determination of X_I fraction according to the COD_t balance [16]

Anaerobic respirometry Most of the methodologies found in literature, as well as the particular ones reviewed here, are based on practical analytical measurements that are often directly related to the ADM1 components, while theoretical assumptions have often been made. Girault et al. [16] propose in their substrate characterization methodology that more attention should be given to biogas production data for ADM1 substrate characterization and to yield the percentage of readily and slowly degradable fractions. This is achieved by a combination of batch experimental degradation tests ("anaerobic respirometry") and numerical interpretation of the methane production rate (MPR) ("optimization of the ADM1 input state variable set"). The proposed method comprises three steps (Fig. 12).

The first step involves the determination of biodegradable COD fractions acetate (S_{ac}), propionate (S_{pro}), butyrate (S_{bu}) and valerate (S_{va}) by high-performance liquid chromatography and an initial COD substrate characterization. Secondly, anaerobic respirometry is applied as an experimental tool to estimate substrate fractionation and the associated degradation kinetics. In principle, the methane production curve gives two distinct phases. The first one represents the percentage of methane that is readily produced from the substrate, hence, related to AA (S_{aa}), monosaccharaides (S_{su}), long chain fatty acids (S_{fa}), valerate (S_{va}), butyrate (S_{bu}), propionate (S_{pro}) and acetate (S_{ac}), while the second one is related to the degradation of proteins (X_{pr}), carbohydrates (X_{ch}) and lipids (X_{li}).

In summary, the studies reviewed here are a proof that the concern of substrate characterization is still of great interest. Since composite materials are complex and variable in their composition (lipids, proteins, carbohydrates and inerts) and also in time, methods are needed that consider these issues with respect to mathematical modeling. Possible approaches could be based upon direct analytical measurements that are implemented straightforwardly into an algorithm for substrate characterization, as shown by Kleerebezem and Van Loosdrecht [17], or indirectly by practical measurements that correlate to components of interest that feed, for instance, a Weender analysis, as shown with the approach by Lübken et al. [31]. Nonetheless, all methods developed for substrate characterization in ADM1 should also aim at keeping the procedure as simple as possible with low workload and on the basis of widely available practical measurements. In this context, the method proposed by Kleerebezem and van Loosdrecht [17] might require the least amount of work with only four measurements, while the ones by Zaher et al. [27], Girault et al. [16] and Lübken et al. [31] will be more laborious when applied. Therefore, the approach by Kleerebezem and van Loosdrecht [17] might also be quite suitable for industrial applications.

All of the studies introduced are based upon COD, which is quite logical as it can be directly correlated to the production of methane in ADM1 (0.35 $m^3 \cdot kg_{COD}^{-1}$). However, this urges modelers to have pure substances or homogenized ones present. This issue was the biggest criticism of Zaher et al. [27] towards the approach by Kleerebezem and van Loosdrecht [17], which mainly suffers from the disadvantage that it assumes the feedstock as a single composite particulate (X_c) with constant composition. Hence, this assumption does not adequately allow for dynamic simulation due to changes in the feedstock composition when dealing with heterogenic wastes. This issue was taken up and resolved by the transformer model of Zaher et al. [27] that applies a CBIM to interface the ADM1 to practical characteristics of an individual waste stream to allow dynamic simulation.

The procedure by Lübken et al. [31] uses analytical analyses, which may not have the accuracy as regular wastewater analyses, because some components are not really analyzed, but are instead roughly calculated upon different fractions. It is, furthermore, limited to lignocellulosic biomass. It proposes splitting the COD_t of a substrate into each input state variable first, while the rest is dedicated to the X_c fraction, which is also recommended by Girault et al. [16]. Drawbacks in Girault and colleagues' methodology are seen when using lower substrate to biomass ratios that allow for a decreased batch test duration (4–10 days) in order to accelerate the procedure, but might also lead to an underestimation of the entire biodegradable fraction and to an overestimation of the non-biodegradable fraction. Moreover, Girault et al. [16] based their waste characterization procedure on pig slurry as substrate, but it still has to be proven to also work for lignocellulosic biomass that shows completely different biodegradability in batch experiments.

Nonetheless, all studies presented here have one common disadvantage, which is related to the protein content. The determination of RP is based upon TKN multiplied by 6.25 (protein-nitrogen conversion factor), which is only valid in the case where the nitrogen content of the protein is 16 %. Kleerebezem and van Loosdrecht [17] used slaughterhouse waste to gather the chemical composition for protein. This can indeed be meaningful, as slaughterhouse waste is characterized by a high solids content that is mainly composed of proteins and fats, with varying amounts of carbohydrates and inorganic compounds [106]. However this class of waste differs in nitrogen content and, hence, in protein. Zaher et al. [27] also used a fictive molecular formula for proteins: $C_6H_{12}O_3N_2$. This underlines the need for a more precise determination of proteins.

All the approaches published have their advantages and disadvantages, and the literature survey does not reveal any trend towards a more or less generally accepted methodology. In case none of the methods introduced can be applied for the simulation study, the following table (Table 7) gives an overview of selected feed fractionations for ADM1 found in literature. As the main challenge for influent fractionation is the allocation of the COD measured to the main ADM1 influent state variables, as described previously, the literature values were normalized to the COD_t and the fractionation is given as a result for the particulate and soluble carbohydrates, proteins, lipids and inerts, as well as for the monosaccharides, AA, long chain fatty acids and VFAs.

6 Conclusions and Future Perspectives

As the state of the art AD models are characterized by a detail-oriented structure, influent fractionation is currently recognized as a severe limitation factor for model application. ADM1 especially needs a substrate characterization, which can hardly be fulfilled by routine analysis. The topic is widely discussed in literature. Several approaches for influent fractionation have already been derived, which are generally based on different concepts. The different nature of these frameworks, e.g. oriented to wastewater and sewage sludge, high solids waste, lignocellulosic biomass or manure-based systems, complicates a comparable analysis of their effectiveness.

Considerable achievements have been made in literature with respect to parameter and model uncertainty analysis. However, and simultaneously astonishing, none of these studies using sophisticated non-linear parameter estimation and model uncertainty procedures include the uncertainty of the influent fractionation methodology within the statistical analysis.

All the influent fractionation approaches presented in literature so far have in common that they are based on offline measurements. Powerful analytical tools are mainly developed to monitor the fermentation process, and most effort is still forced in this direction. These tools are incontrovertibly important for model validation procedures, but, as variations of influent composition can be highly dynamic, high-capacity analytical methods are here similarly required. To account for this, modern online monitoring tools allowing substrate characterization, such as NIRS, should also be calibrated to the demands of sophisticated AD models.

Table 7 Calculated distribution of the influent COD_t to ADM1 influent state variables for various substrates based on literature data

Substrate	X_{ch} (%)	X_{pr} (%)	X_{li} (%)	X_i (%)	S_{su} (%)	S_{aa} (%)	S_{fa} (%)	S_{VFA} (%)	S_i (%)	References
Apple pulp	0.256	0.011	0.055	0.255					0.422	[107]
Blackwater	0.114	0.114	0.146	0.201			0.213	0.213		[108]
Corn processing wastewater	0.770	0.060	0.020	0.050					0.100	[109]
Grass silage (heterofermentatively treated)	0.488	0.149	0.015	0.314				0.010	0.025	[32]
Grass silage	0.401	0.187	0.033	0.379						[33]
Microalgae	0.072	0.360	0.198	0.270	0.100					[110]
Olive mill solid waste	0.359	0.076	0.103	0.462						[50]
Olive mill wastewater	0.099	0.118	0.124	0.125	0.161	0.056	0.087	0.136	0.093	[50]
Olive pulp	0.088	0.080	0.413	0.203	0.093		0.047		0.078	[111]
Opium alkaloid	0.034		0.012	0.024	0.296	0.181	0.106	0.330	0.016	[112]
Orange pulp	0.477	0.020	0.014	0.337					0.153	[107]
Pear pulp	0.399	0.016	0.084	0.134					0.367	[107]
Pig manure	0.461	0.202	0.161	0.033					0.143	[107]
Pig slurry	0.102	0.018	0.034	0.430	0.181	0.033	0.060	0.141		[16]
Rape	0.509	0.115	0.112	0.152					0.112	[107]
Sunflower	0.506	0.198	0.034	0.078					0.184	[107]
Swine manure	0.036	0.164	0.191	0.072		0.111	0.124	0.276	0.026	[113]
Waste activated sludge	0.096	0.160	0.138	0.606						[114]
Waste activated sludge	0.098	0.162	0.057	0.451	0.072	0.118	0.042			[16]
Waste activated sludge (pretreated at 165 °C)	0.135	0.149	0.242	0.475						[114]
Wine distillery wastewater	0.073				0.443			0.484		[115]

Values were normalized for the COD_t

Quantification of active biomass within the influent is challenging. No reliable methodology has been published so far. While this topic is of minor importance for continuous fermentation systems, the inoculum of batch assays, from which kinetic parameter studies are commonly performed, can only inadequately be characterized.

The ADM1 protein fraction in most applications is indirectly determined by Kjeldahl analysis. The total nitrogen gathered is multiplied by the factor 6.25 to obtain the overall protein concentration or content. The factor is equivalent to the reciprocal of the average nitrogen content of RP that is assumed to be 16 %. If this is not the case, errors have to be logically expected. Moreover, protein determination is a very time-consuming technique to carry out. Several new substrates have been tested for energy recovery in AD systems in recent years. Some of them, such as algae, are mainly composed of proteins, leading to high biogas production rates. A more detailed characterization of the AA composition of proteins is essential for these substrates. This topic has been rather neglected in literature.

The main challenge for future work will be the consolidation of the vast numbers of methodologies and approaches presented in literature. It is evident that a common and scientifically accepted guideline for influent fractionation will further assist and stimulate the application of structured models such as the ADM1.

References

1. Andrews JF (1996) Dynamic model of the anaerobic digestion process. ASCE J Sanit Eng Div 95:95–116
2. Husain A (1998) Mathematical models of the kinetics of anaerobic digestion—A selected review. Biomass Bioenergy 14:561–571
3. Saravanan V, Sreekrishnan TR (2006) Modelling anaerobic biofilm reactors—A review. J Environ Manage 81:1–18
4. Donoso-Bravo A, Mailier J, Martin C, Rodríguez J, Aceves-Lara CA, Vande Wouwer A (2011) Model selection, identification and validation in anaerobic digestion: a review. Water Res 45:5347–5364
5. Tomei MC, Braguglia CM, Cento G, Mininni G (2009) Modeling of anaerobic digestion of sludge. Crit Rev Environ Sci Technol 39:1003–1051
6. Lübken M, Gehring T, Wichern M (2010) Microbiological fermentation of lignocellulosic biomass: current state and prospects of mathematical modeling. Appl Microbiol Biotechnol 85:1643–1652
7. Batstone DJ (2006) Mathematical modelling of anaerobic reactors treating domestic wastewater: Rational criteria for model use. Rev Environ Sci Biotechnol 5:57–71
8. Batstone DJ, Keller J, Angelidaki I, Kalyuzhnyi SV, Pavlostathis SG, Rozzi A, Sanders WTM, Siegrist H, Vavilin VA (2002) The IWA Anaerobic Digestion Model No 1 (ADM1). Water Sci Technol 45:65–73
9. Batstone DJ, Keller J, Angelidaki I, Kalyuzhnyi SV, Pavlostathis SG, Rozzi A, Sanders WTM, Siegrist H, Vavilin VA (2002) Anaerobic digestion model no. 1 (ADM1). IWA Publishing, London, UK
10. Holm-Nielsen JB, Dahl CK, Esbensen KH (2006) Representative sampling for process analytical characterization of heterogeneous bioslurry systems—a reference study of sampling issues in PAT. Chemom Intell Lab Syst 83:114–126

11. García-Gen S, Lema JM, Rodríguez J (2013) Generalised modelling approach for anaerobic co-digestion of fermentable substrates. Bioresour Technol 147:525–533
12. Fezzani B, Ben Cheikh R (2009) Extension of the anaerobic digestion model No. 1 (ADM1) to include phenol compounds biodegradation processes for simulating the anaerobic co-digestion of olive mill wastes at mesophilic temperature. J Hazard Mater 172:1430–1438
13. Fedorovich V, Lens P, Kalyuzhnyi S (2003) Extension of anaerobic digestion model no. 1 with processes of sulfate reduction. Appl Biochem Biotechnol 109:33–45
14. Mata-Alvarez J, Macé S, Llabrés P (2000) Anaerobic digestion of organic solid wastes. An overview of research achievements and perspectives. Bioresour Technol 74:3–16
15. Lübken M, Koch K, Gehring T, Horn H, Wichern M (2015) Parameter estimation and long-term process simulation of a biogas reactor operated under trace elements limitation. Appl Energy 142:352–360
16. Girault R, Bridoux G, Nauleau F, Poullain C, Buffet J, Steyer JP, Sadowski AG, Béline F (2012) A waste characterisation procedure for ADM1 implementation based on degradation kinetics. Water Res 46:4099–110
17. Kleerebezem R, Van Loosdrecht MCM (2006) Waste characterization for implementation in ADM1. Water Sci Technol 54:167–174
18. Zaher U, Chen S (2006) Interfacing the IWA anaerobic digestion model no 1 (ADM1) with manure and solid waste characteristics. Water Environ Found 1:3162–3175
19. Huete E, de Gracia M, Ayesa E, Garcia-Heras JL (2006) ADM1-based methodology for the characterisation of the influent sludge in anaerobic reactors. Water Sci Technol 54:157–166
20. Boe K, Steyer JP, Angelidaki I (2008) Monitoring and control of the biogas process based on propionate concentration using online VFA measurement. Water Sci Technol 57:661–666
21. Boe K, Batstone DJ, Steyer JP, Angelidaki I (2010) State indicators for monitoring the anaerobic digestion process. Water Res 44:5973–5980
22. Nacke T, Brückner K, Göller A, Kaufhold S, Nakos X, Noack S, Stöber H, Beckmann D (2005) New type of dry substances content meter using microwaves for application in biogas plants. Anal Bioanal Chem 383:752–757
23. Triolo JM, Ward AJ, Pedersen L, Løkke MM, Qu H, Sommer SG (2014) Near Infrared Reflectance Spectroscopy (NIRS) for rapid determination of biochemical methane potential of plant biomass. Appl Energy 116:52–57
24. Jacobi HF, Ohl S, Thiessen E, Hartung E (2012) NIRS-aided monitoring and prediction of biogas yields from maize silage at a full-scale biogas plant applying lumped kinetics. Bioresour Technol 103:162–172
25. Copp JB, Jeppsson U, Rosen C (2003) Towards an ASM1—ADM1 state variable interface for plant-wide wastewater treatment modeling. In: Proceedings of the 76th annual WEF conference Expo pp 498–510
26. Wett B, Eladawy A, Ogurek M (2006) Description of nitrogen incorporation and release in ADM1. Water Sci Technol 54:67–76
27. Zaher U, Buffiere P, Steyer JP, Chen S (2009) A procedure to estimate proximate analysis of mixed organic wastes. Water Environ Res 81:407–415
28. Nopens I, Batstone DJ, Copp JB, Jeppsson U, Volcke E, Alex J, Vanrolleghem PA (2009) An ASM/ADM model interface for dynamic plant-wide simulation. Water Res 43:1913–1923
29. Yasui H, Goel R, Li YY, Noike T (2008) Modified ADM1 structure for modelling municipal primary sludge hydrolysis. Water Res 42:249–59
30. Girault R, Steyer JP, Zaher U, Sadowski AG, Nopens I, Béline F, Kujawski O, Holm NC, Rönner-Holm SGE (2010) Influent fractionation and parameter calibration for ADM1 : lab-scale and full-scale experiments. Wastewater treatment modelling seminar, Mont-Sainte-Anne, Canada, March 28–30
31. Lübken M, Wichern M, Schlattmann M, Gronauer A, Horn H (2007) Modelling the energy balance of an anaerobic digester fed with cattle manure and renewable energy crops. Water Res 41:4085–4096

32. Wichern M, Gehring T, Fischer K, Andrade D, Lübken M, Koch K, Gronauer A, Horn H (2009) Monofermentation of grass silage under mesophilic conditions: Measurements and mathematical modeling with ADM 1. Bioresour Technol 100:1675–1681
33. Koch K, Lübken M, Gehring T, Wichern M, Horn H (2010) Biogas from grass silage—Measurements and modeling with ADM1. Bioresour Technol 101:8158–8165
34. Gossett JM, Belser RL (1982) Anaerobic digestion of waste activated sludge. J Environ Eng Div ASCE 108:1101–1120
35. Kampas P, Parsons SA, Pearce P, Ledoux S, Vale P, Churchley J, Cartmell E (2007) Mechanical sludge disintegration for the production of carbon source for biological nutrient removal. Water Res 41:1734–42
36. Kim M, Ahn YH, Speece R (2002) Comparative process stability and efficiency of anaerobic digestion; mesophilic vs. thermophilic. Water Res 36:4369–4385
37. Parravicini V, Svardal K, Hornek R, Kroiss H (2008) Aeration of anaerobically digested sewage sludge for COD and nitrogen removal: optimization at large-scale. Water Sci Technol 57:257–264
38. Münch E, Keller J, Lant P, Newell R (1999) Mathematical modelling of prefermenters—I. Model development and verification. Water Res 33:2757–2768
39. Rice EW, Baird RB, Eaton AD, Clesceri LS (2012) Standard methods for the examination of water and wastewater. American Public Health Association, Washington DC
40. VDI 4630 (2006) Fermentation of organic materials—Characterisation of the substrate, sampling, collection of material data, fermentation tests. Verein Deutscher Ingenieure, Düsseldorf, Germany
41. Kreuger E, Nges IA, Björnsson L (2011) Ensiling of crops for biogas production: effects on methane yield and total solids determination. Biotechnol Biofuels 4:44
42. Ekama GA (2009) Using bioprocess stoichiometry to build a plant-wide mass balance based steady-state WWTP model. Water Res 43:2101–2120
43. Koch K, Drewes JE (2014) Alternative approach to estimate the hydrolysis rate constant of particulate material from batch data. Appl Energy 120:11–15
44. Angelidaki I, Sanders W (2004) Assessment of the anaerobic biodegradability of macropollutants. Rev Environ Sci Bio/Technology 3:117–129
45. Jimenez J, Gonidec E, Cacho Rivero JA, Latrille E, Vedrenne F, Steyer JP (2014) Prediction of anaerobic biodegradability and bioaccessibility of municipal sludge by coupling sequential extractions with fluorescence spectroscopy: towards ADM1 variables characterization. Water Res 50:359–372
46. Lesteur M, Bellon-Maurel V, Gonzalez C, Latrille E, Roger JM, Junqua G, Steyer JP (2010) Alternative methods for determining anaerobic biodegradability: a review. Process Biochem 45:431–440
47. Ekama GA, Dold PL (1986) Procedures for determining influent cod fractions and the maximum specific growth rate of heterotrophs in activated sludge systems. Water Sci Technol 18:91–114
48. Wichern M, Lübken M, Blömer R, Rosenwinkel KH (2003) Efficiency of the Activated Sludge Model no. 3 for German wastewater on six different WWTPs. Water Sci Technol 47:211–218
49. Blumensaat F, Keller J (2005) Modelling of two-stage anaerobic digestion using the IWA anaerobic digestion model no. 1 (ADM1). Water Res 39:171–183
50. Boubaker F, Ridha BC (2008) Modelling of the mesophilic anaerobic co-digestion of olive mill wastewater with olive mill solid waste using anaerobic digestion model no. 1 (ADM1). Bioresour Technol 99:6565–6577
51. Lettinga G, Hulshoff Pol LW (1991) UASB-Process design for various types of wastewaters. Water Sci Technol 24:87–107
52. Hinken L, Huber M, Weichgrebe D, Rosenwinkel KH (2014) Modified ADM1 for modelling an UASB reactor laboratory plant treating starch wastewater and synthetic substrate load tests. Water Res 64:82–93

53. Healy JB, Young LY (1979) Anaerobic biodegradation of eleven aromatic compounds to methane. Appl Environ Microbiol 38:84–89
54. Heider J, Fuchs G (1997) Anaerobic metabolism of aromatic compounds. Eur J Biochem 243:577–596
55. Hongwei Y, Zhanpeng J, Shaoqi S (2004) Anaerobic biodegradability of aliphatic compounds and their quantitative structure biodegradability relationship. Sci Total Environ 322:209–219
56. Chen Y, Cheng JJ, Creamer KS (2008) Inhibition of anaerobic digestion process: a review. Bioresour Technol 99:4044–4064
57. Delgadillo-Mirquez L, Lardon L, Steyer JP, Patureau D (2011) A new dynamic model for bioavailability and cometabolism of micropollutants during anaerobic digestion. Water Res 45:4511–4121
58. Naumann C, Bassler R (2004) Handbuch der landwirtschaftlichen Versuchs- und Untersuchungsmethodik: (Methodenbuch), vol 3. neu bearbeitete Auflage, VDLUFA-Verlag, Darmstadt, Germany
59. Van Soest PJ, Wine RH (1967) Use of detergents in the analysis of fibrous feeds. IV. Determination of plant cell-wall constituents. J Assoc Off Anal Chem 50–59
60. Grieder C, Mittweg G, Dhillon BS, Montes JM, Orsini E, Melchinger AE (2012) Kinetics of methane fermentation yield in biogas reactors: Genetic variation and association with chemical composition in maize. Biomass Bioenergy 37:132–141
61. Schönberg M, Linke B (2012) The influence of the temperature regime on the formation of methane in a two-phase anaerobic digestion process. Eng Life Sci 12:279–286
62. Amon T, Amon B, Kryvoruchko V, Zollitsch W, Mayer K, Gruber L (2007) Biogas production from maize and dairy cattle manure—Influence of biomass composition on the methane yield. Agric Ecosyst Environ 118:173–182
63. Wichern M, Lübken M, Horn H, Schlattmann M, Gronauer A (2008) Investigations and mathematical simulation on decentralized anaerobic treatment of agricultural substrate from livestock farming. Water Sci Technol 58:67–72
64. Biernacki P, Steinigeweg S, Borchert A, Uhlenhut F, Brehm A (2013) Application of anaerobic digestion model no. 1 for describing an existing biogas power plant. Biomass Bioenergy 59:441–447
65. Biernacki P, Steinigeweg S, Borchert A, Uhlenhut F (2013) Application of anaerobic digestion model no. 1 for describing anaerobic digestion of grass, maize, green weed silage, and industrial glycerine. Bioresour Technol 127:188–94
66. Myint M, Nirmalakhandan N, Speece RE (2007) Anaerobic fermentation of cattle manure: modeling of hydrolysis and acidogenesis. Water Res 41:323–332
67. Orhon D, Çokgör EU, Sözen S (1998) Dual hydrolysis model of the slowly biodegradable substrate in activated sludge systems. Biotechnol Tech 12:737–741
68. Fernandez I, Mahieu N, Cadisch G (2003) Carbon isotopic fractionation during decomposition of plant materials of different quality. Global Biogeochem Cycles 17:1–9
69. Gehring T, Klang J, Niedermayr A, Berzio S, Immenhauser A, Klocke M, Wichern M, Lübken M (2015) Determination of methanogenic pathways through carbon isotope (δ13C) analysis for the two-stage anaerobic digestion of high-solids substrates. Environ Sci Technol. doi:10.1021/es505665z
70. Koch K, Wichern M, Lübken M, Horn H (2009) Mono fermentation of grass silage by means of loop reactors. Bioresour Technol 100:5934–5940
71. Batstone DJ, Tait S, Starrenburg D (2009) Estimation of hydrolysis parameters in full-scale anerobic digesters. Biotechnol Bioeng 102:1513–1520
72. Madigan MT (2012) Brock biology of microorganisms, 13th edn. International Microbiology. doi:10.1016/B978-1-4832-3136-5.50010-3
73. Henze M, Van Loosdrecht MCM, Ekama GA, Brdjanovic D (2008) Biological wastewater treatment : principles, modelling and design. IWA Publishing, London
74. Fan K, Wang W (2003) What is the minimum number of letters required to fold a protein? J Mol Biol 328:921–926

75. Nanni L, Brahnam S, Lumini A (2014) Prediction of protein structure classes by incorporating different protein descriptors into general Chou's pseudo amino acid composition. J Theor Biol 360:109–116
76. Lowry OH, Rosebrough NJ, Farr AL, Randall RJ (1951) Protein measurement with the Folin phenol reagent. J Biol Chem 193:265–275
77. Luque de Castro MD, Priego-Capote F (2010) Soxhlet extraction: past and present panacea. J Chromatogr A 1217:2383–2389
78. Luque de Castro MD, García-Ayuso LE (1998) Soxhlet extraction of solid materials: an outdated technique with a promising innovative future. Anal Chim Acta 369:1–10
79. Manirakiza P, Covaci A, Schepens P (2001) Comparative study on total lipid determination using Soxhlet, Roese-Gottlieb, Bligh & Dyer, and Modified Bligh & Dyer extraction methods. J Food Compos Anal 14:93–100
80. Yemm EW, Willis AJ (1954) The estimation of carbohydrates in plant extracts by anthrone. Biochem J 57:508–514
81. Updegraff DM (1969) Semimicro determination of cellulose in biological materials. Anal Biochem 32:420–424
82. Bauer S, Ibáñez AB (2014) Rapid determination of cellulose. Biotechnol Bioeng 111:2355–2357
83. Eckenfelder WW, Weston RF (1956) Kinetics of biological oxidation. In: McCabe BJ, Eckenfelder WW (eds) Biological treatment of sewage and industrial wastes, vol 1. Reinhold Publ. Co., New York, USA, pp 18–34
84. Servizi JA, Bogan RH (1963) Free energy as a parameter in biological treatment. J San Eng Div, ASCE 63:17–40
85. Kleerebezem R, van Loosdrecht MCM (2006) Critical analysis of some concepts proposed in ADM1. Water Sci Technol 54:51–57
86. Rodríguez J, Lema JM, van Loosdrecht MCM, Kleerebezem R (2006) Variable stoichiometry with thermodynamic control in ADM1. Water Sci Technol 54:101–110
87. Penumathsa BKV, Premier GC, Kyazze G, Dinsdale R, Guwy AJ, Esteves S, Rodríguez J (2008) ADM1 can be applied to continuous bio-hydrogen production using a variable stoichiometry approach. Water Res 42:4379–4385
88. Wallace JC, Lopes MA, Paiva E, Larkins BA (1990) New methods for extraction and quantitation of Zeins reveal a high content of gamma-Zein in modified opaque-2 maize. Plant Physiol 92:191–196
89. Dennstedt M, Haßler F (1906) Über den Abbau von Eiweiß. Zeitschrift für Physiol Chemie 48:489–504
90. Ramsay IR, Pullammanappallil PC (2001) Protein degradation during anaerobic wastewater treatment: derivation of stoichiometry. Biodegradation 12:247–256
91. Flotats X, Palatsi J, Ahring BK, Angelidaki I (2006) Identifiability study of the proteins degradation model, based on ADM1, using simultaneous batch experiments. Water Sci Technol 54:31–39
92. Heger J, Salek M, Eggum BO (1990) Nutritional value of some Czechoslovak varieties of wheat, triticale and rye. Anim Feed Sci Technol 29:89–100
93. Becker EW (2007) Micro-algae as a source of protein. Biotechnol Adv 25:207–210
94. Yuan XZ, Shi XS, Yuan CX, Wang YP, Qiu YL, Guo RB, Wang LS (2014) Modeling anaerobic digestion of blue algae: stoichiometric coefficients of amino acids acidogenesis and thermodynamics analysis. Water Res 49:113–123
95. Mu TH, Tan SS, Xue YL (2009) The amino acid composition, solubility and emulsifying properties of sweet potato protein. Food Chem 112:1002–1005
96. Angelidaki I, Ahring BK (1994) Anaerobic thermophilic digestion of manure at different ammonia loads: effect of temperature. Water Res 28:727–731
97. Lübken M, Wichern M, Letsiou I, Kehl O, Bischof F, Horn H (2007) Thermophilic anaerobic digestion in compact systems: investigations by modern microbiological techniques and mathematical simulation. Water Sci Technol 56:19–28

98. Lübken M, Wichern M, Bischof F, Prechtl S, Horn H (2007) Development of an empirical mathematical model for describing and optimizing the hygiene potential of a thermophilic anaerobic bioreactor treating faeces. Water Sci Technol 55:95–102
99. Girault R, Rousseau P, Steyer JP, Bernet N, Béline F (2011) Combination of batch experiments with continuous reactor data for ADM1 calibration: application to anaerobic digestion of pig slurry. Water Sci Technol 63:2575–2582
100. Lebuhn M, Munk B, Effenberger M (2014) Agricultural biogas production in Germany—from practice to microbiology basics. Energy Sustain Soc 4:10
101. Metzker ML (2010) Sequencing technologies—the next generation. Nat Rev Genet 11:31–46
102. Loehr RC (1974) Agricultural waste management—problems, processes, and approaches. Academic Press, New York, 576 pp
103. Mosey F (1983) Mathematical modeling of the anaerobic digestion process: Regulatory mechanisms for the formation of short chain volatile acids from glucose. Water Sci Technol 15:209–232
104. Kleerebezem R, Van Loosdrecht MCM (2010) A generalized method for thermodynamic state analysis of environmental systems. Crit Rev Environ Sci Technol 40:1–54
105. Vanrolleghem PA, Rosen C, Zaher U, Copp J, Benedetti L, Ayesa E, Jeppsson U (2005) Continuity-based interfacing of models for wastewater systems described by Petersen matrices. Water Sci Technol 52:493–500
106. Ben M, Kennes C, Veiga MC (2013) Air pollution prevention and control: Bioreactors and bioenergy. In: Kennes C, Veiga MC (eds) Wiley-Blackwell, Chichester, 570 pp
107. Galí A, Benabdallah T, Astals S, Mata-Alvarez J (2009) Modified version of ADM1 model for agro-waste application. Bioresour Technol 100:2783–2790
108. Feng Y, Behrendt J, Wendland C, Otterpohl R (2006) Parameter analysis of the IWA anaerobic digestion model no. 1 for the anaerobic digestion of blackwater with kitchen refuse. Water Sci Technol 54:139–147
109. Ersahin ME, Insel G, Dereli RK, Ozturk I, Kinaci C (2007) Model based evaluation for the anaerobic treatment of corn processing wastewaters. CLEAN–Soil. Air, Water 35:576–581
110. Mairet F, Bernard O, Ras M, Lardon L, Steyer JP (2011) Modeling anaerobic digestion of microalgae using ADM1. Bioresour Technol 102:6823–6829
111. Kalfas H, Skiadas IV, Gavala HN, Stamatelatou K, Lyberatos G (2006) Application of ADM1 for the simulation of anaerobic digestion of olive pulp under mesophilic and thermophilic conditions. Water Sci Technol 54:149–156
112. Dereli RK, Ersahin ME, Ozgun H, Ozturk I, Aydin AF (2010) Applicability of anaerobic digestion model no. 1 (ADM1) for a specific industrial wastewater: opium alkaloid effluents. Chem Eng J 165:89–94
113. Jurado E, Gavala HN, Skiadas I (2012) ADM1-based modeling of anaerobic digestion of swine manure fibers pretreated with aqueous ammonia soaking. In: Proceedings of the 4th international symposium on energy from biomass and waste, San Servolo, Venice, Italy, 12–15 Nov 2012
114. Ramirez I, Mottet A, Carrère H, Déléris S, Vedrenne F, Steyer JP (2009) Modified ADM1 disintegration/hydrolysis structures for modeling batch thermophilic anaerobic digestion of thermally pretreated waste activated sludge. Water Res 43:3479–3492
115. Ramirez I, Steyer JP (2008) Modeling microbial diversity in anaerobic digestion. Water Sci Technol 57:265–270

Fate of Trace Metals in Anaerobic Digestion

F.G. Fermoso, E.D. van Hullebusch, G. Guibaud, G. Collins, B.H. Svensson, C. Carliell-Marquet, J.P.M. Vink, G. Esposito and L. Frunzo

Abstract A challenging, and largely uncharted, area of research in the field of anaerobic digestion science and technology is in understanding the roles of trace metals in enabling biogas production. This is a major knowledge gap and a multifaceted problem involving metal chemistry; physical interactions of metal and solids; microbiology; and technology optimization. Moreover, the fate of trace metals, and the chemical speciation and transport of trace metals in environments—often agricultural lands receiving discharge waters from anaerobic digestion processes—simultaneously represents challenges for environmental protection and opportunities to close process loops in anaerobic digestion.

Keywords Metal speciation · Trace metal microbiology · Bioavailability · Anaerobic digestion · Biogas · Mathematical modeling

Contents

F.G. Fermoso (✉)
Instituto de La Grasa, C.S.I.C., Campus Pablo de Olavide, Ctra. de Utrera Km.1, 41013 Seville, Spain
e-mail: fgfermoso@ig.csic.es

E.D. van Hullebusch
Université Paris-Est, Laboratoire Géomatériaux et Environnement (EA 4508), UPEM, 77454 Marne-La-Vallée, France

G. Guibaud
Faculté des Sciences et Techniques, University of Limoges Groupement de Recherche, Eau, Sol, Environnement, 123 Av. Albert Thomas, 87 060 Limoges, France

G.M. Guebitz et al. (eds.), *Biogas Science and Technology*,
Advances in Biochemical Engineering/Biotechnology 151,
DOI 10.1007/978-3-319-21993-6_7

1 Introduction

Anaerobic digestion (AD) is an attractive technology for the production of bioenergy [1] and for wastewater treatment [2, 3]. The advantages of AD for waste treatment include: production of versatile and storable fuel (biogas/methane); potential of high organic loading; reduced carbon footprint; and suitability for integration into a wide variety of process configurations and scales [2, 4]. AD is increasingly applied world-wide; in 2011, 8760 anaerobic bioreactors were reported in Europe, most of which were used to produce renewable energy from organic feedstock [5, 6], itself a limited resource. Internationally, the research efforts to maximize biogas yields has increased ten-fold over the past decade [7, 8]. Still, a critical research question remains open: How does trace metal availability limit

G. Collins
School of Natural Sciences, National University of Ireland Galway, Microbial Ecophysiology and EcoEngineering Laboratory, University Road, Galway, Ireland

G. Collins
School of Engineering, University of Glasgow, Rankine Building, Oakfield Avenue, Glasgow G12 8LT, UK

B.H. Svensson
Department Thematic Studies Water and Environment, Linköping University, SE-581 83 Linköping, Linköping, Sweden

C. Carliell-Marquet
School of Civil Engineering, College of Engineering and Physical Sciences, University of Birmingham, Birmingham B15 2TT, UK

J.P.M. Vink
Deltares Foundation, Princetonlaan 6, 3508 Utrecht, AL, The Netherlands

G. Esposito
Department of Civil and Mechanical Engineering, University of Cassino and the Southern Lazio, Via Di Biasio, 43, 03043 Cassino, FR, Italy

L. Frunzo
Department of Mathematics and Applications Renato Caccioppoli, University of Naples Federico II, Via Claudio 21, 80125 Naples, Italy

biogas production yield and how can this limitation be circumvented? To answer this question with confidence, in such a complex matrix of substrates, microorganisms and chemical products, requires spanning fundamental molecular sciences to engineering applications; only then will the underpinning science be coupled successfully to engineering-led systems, benefitting end-users and producing renewable energy more efficiently.

Both the beneficial, and inhibitory or toxic effects of trace metals (TM) in anaerobic treatment processes have long been an interesting topic for researchers. Chalcogens—which include the trace elements selenium (Se), tellurium (Te) and radioactive polonium (Po)—although not metals, are included in the term TM in this chapter.

The roles of TM in anaerobic processes are significant. Anaerobic fermentation and microbial growth is dependent on the availability and/or optimal supply of nutrients. Free metal ion availability is an important parameter that should be considered [9]. The requirements of various methanogens for iron (Fe), nickel (Ni), cobalt (Co), molybdenum (Mo), selenium (Se) and tungsten (W) have already been reported [10, 11]. Furthermore, the effects of TM such as Fe, Ni, Co, Zn, Mo and Cu on anaerobic treatment of various types of industrial effluents have also been investigated in detail [12]. On the other hand, TM requirements of agricultural biogas systems operated with solid organic matter, such as energy crops, animal excreta, crop residues and the organic fraction of municipal solid wastes (OFMSW), are seldom reported in the literature, despite the exponentially-increasing interest in biogas production from renewable sources [13].

Since industrial wastewaters and sludges contain elevated amounts of TM, most of the research activity has obviously been directed towards investigating the inhibitory and toxic effects of these elements on anaerobic wastewater treatment bioprocesses. During anaerobic conversion of energy crops, animal excreta, crop residues, OFMSW or any other type of organic wastes (e.g. food wastes) to biogas-methane, the availability, or lack, of TM, such as Fe, Co, Ni, Zn, Mo, W and Se, plays a significant role in maintaining stable and efficient conversion processes. Recently, it has been shown that food waste appears deficient in some TM, such as Co and Se, required by the anaerobic digestion process when operating at high ammonia concentrations [14, 15]. Such findings have major implications for successful and expanded application of AD biotechnology.

The area of TM bioavailability in AD systems calls for studies from several perspectives. Thus, efforts from a multitude of research fields are needed and the area is, by necessity, interdisciplinary. TM research draws on: physics e.g. rheology, diffusion, adsorption; organic, inorganic, physical and analytical chemistry; microbial physiology, genetics and biotechnology (laboratory-, pilot- and full-scale systems); agronomy and forestry biofertilization, among others. This review focuses on five areas covering the fate and roles of TM in anaerobic biotechnologies: TM speciation and bioavailability in AD environments; TM microbiology and microbial ecology; biotechnology; fate of TM in the environment; and mathematical modelling.

2 Trace Metals Speciation, Fractioning and Bioavailability

The total metal concentration, the physico-chemicals conditions during digestion (pH and redox potential), and the reaction kinetics of chemical processes occurring both in liquid (i.e. trace elements reduction, precipitation or complexation) and solid phases (i.e. trace elements sorption) play key roles influencing the chemical speciation of TM in AD bioreactors. For instance, an increase of the pH value results in a decrease of the solubility of cationic metals in the matrix. The precipitation of metals by sulfide (S^{2-}), carbonate ($CO_3{}^{2-}$) and phosphate ($PO_4{}^{3-}$), and their deposition in the bioreactor sludges/biofilms, plays an important role in nutrients, and TM, turnover [12, 16–18].

For instance, the presence of sulfide in biogas reactors affects the availability of TM needed for growth and metabolic activities of the microorganisms involved in AD. The solubility constants for complexes between most of the essential TM and sulfides are low and may lead to reduced bioavailability for these compounds [19]. High sulfide content in AD bioreactors may, therefore, result in suboptimal biogas production [20] and lead to a shift of micronutrients away from soluble bioavailable forms toward the bioinactive complex or precipitate species during AD [21]. However, due to its high corrosive properties, hydrogen sulfide affects biogas plant equipment and downstream biogas utilization devices, e.g. pipes and motors for converting the biogas to electricity and heat. Therefore, Fe in form of $FeCl_2$ is sometimes dosed to biogas reactors to precipitate the sulfide formed and to avoid high concentration of hydrogen sulfide in the biogas [18, 20, 22]. This leads to a complete change of the sulfur turnover in AD reactors (Fig. 1) and also to a significant change of macro and micronutrient speciation. There is a strong interaction of added Fe and the micronutrients in the matrix: micronutrients may react with the Fe-sulfide releasing Fe^{2+}. The resulting Fe^{2+} may form precipitates as phosphates ($Fe_3(PO_4)_2$) or carbonates ($FeCO_3$) [18]. Consequently, bioreactors have a considerable ability to sequester Fe^{2+}-ions in the sludge. Simultaneously, non-alkali metals (e.g. Ca^{2+}, Mg^{2+}) form soluble ion pairs with a number of anions: $HCO_3{}^-$, $CO_3{}^{2-}$, OH^-, $SO_4{}^{2-}$, S^{2-} [17].

Furthermore, TM can interact with organic substances present in the bioreactor, such as the microbial cell wall (e.g. [23]), Extracellular Polymeric Substances (EPS) (e.g. [24]), Soluble Microbial Products (SMP) (e.g. [25]), organic substrates (e.g. yeast extract, [26]), organic sulfur compounds (e.g. organic thiol compounds (RS^-), [22]), synthetic complexing agents, such as EDTA (e.g. [27, 28]) or organic acids (e.g. [25]). The main interactions between organic substances and TM occur due to the presence ionisable functional groups, such as carboxylic, phosphoric, amino, and hydroxylic groups [29].

Molecular-level characterization of dissolved organic matter in biogas bioreactors digesting different types of substrates by electrospray ionization Fourier transform ion cyclotron resonance mass spectrometry (ESI-FT-ICR-MS) revealed that S-containing dissolved organic compounds including thiols may contribute up to 30 % of dissolved organic molecules [30]. In addition, during AD of dewatered

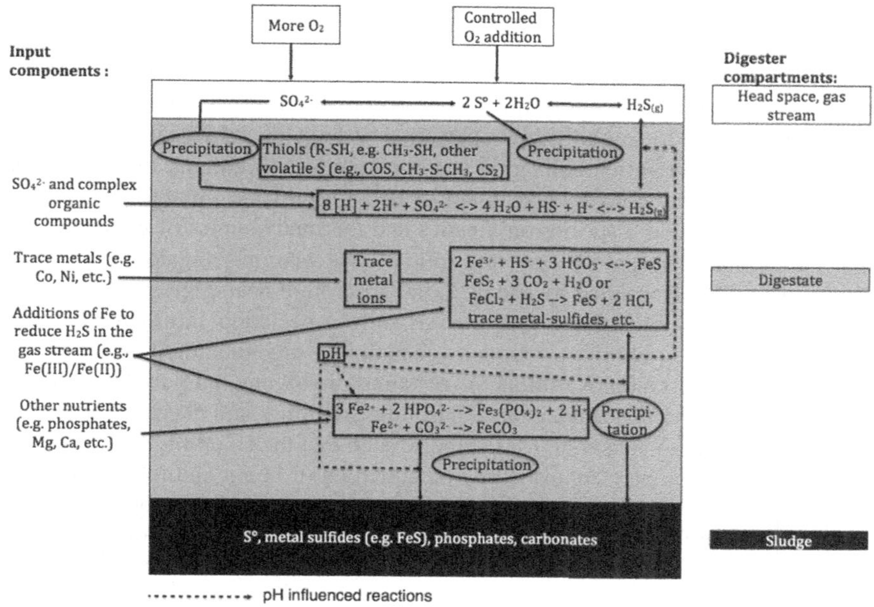

Fig. 1 Sulfur turnover in biogas bioreactors and its influence on TM and nutrients speciation [modified from 17]

sewage sludge, Li et al. [31] demonstrated chemical changes of dissolved organic matter using PARAllel FACtor analysis for decomposing fluorescence excitation emission matrices (EEM-PARAFAC) and two-dimensional FTIR correlation spectroscopy. Li et al. [31] showed that fluorescence intensities of the components relating to tyrosine-like, tryptophan-like and humic-like groups increased in Dissolved Organic Matter, implying that these groups were reluctant to biodegrade and could participate in TM complexation. It has been recently shown that trace elements may be significantly complexed by organic compounds harbouring thiol functional groups [22].

Complexation reactions (in the liquid phase or the solid phase) play an important role in bioreactors making a particular TM either more or less bioavailable. The level of soluble TM in the presence of CO_3^{2-} and S^{2-} may be increased by a factor of up to 10^4 by organic complexation, avoiding precipitation as carbonates or sulfides [19]. However, up to now little quantitative information has been published regarding the contribution of organic substances in TM complexation and its effect on TM bioavailability.

AD bioreactors where sulfide concentration is high and where iron dosing is needed to prevent corrosion problems constitute good models for studying TM bioavailability. In first approach, the chemical forms and potential bioavailability of trace elements could be easily examined by sequential extraction, acid volatile sulfide extraction (AVS) and simultaneously extracted metals analysis [32].

Furthermore, knowing the importance of sulfur in trace elements bioavailability and speciation, sulfur speciation in solid phase could be also examined by X-ray absorption near edge structure spectroscopy at Sulfur K-edge [18, 20, 22, 33]. For instance, Shakeri Yekta et al. [22] assessed major chemical reactions and chemical forms contributing to solubility and speciation of Fe, Co, and Ni during anaerobic digestion of sulfur (S)-rich stillage in semi-continuous stirred tank biogas reactors (SCSTR). A particular focus was given to the study of the influence of reduced inorganic and organic S species on kinetics and thermodynamics of the metals and their partitioning between aqueous and solid phases were investigated. Solid phase S speciation was determined by use of X-ray absorption near-edge spectroscopy at S K-edge. By combining the quantitative sulfur speciation information with a thermodynamic equilibrium model including precipitation/dissolution of metal-sulfides and complex formation reactions involving inorganic sulfide and organic thiols as the major reactions controlling solubility and chemical speciation of the TM, Shakeri Yekta et al. [22] demonstrated that the solubility and speciation of supplemented Fe were controlled by precipitation of $FeS_{(s)}$ and formation of the aqueous complexes of Fe-sulfide and Fe-thiol. The relatively high solubility of Co ($\sim$20 % of total Co content) was attributed to the formation of compounds other than Co-sulfide and Co-thiol, presumably of microbial origin. Ni had lower solubility than Co and its speciation was regulated by interactions with $FeS_{(s)}$ (e.g. co-precipitation, adsorption, and ion substitution) in addition to precipitation/dissolution of discrete $NiS_{(s)}$ phase and formation of aqueous Ni-sulfide complexes. Such analytical approaches could be successfully implemented provided that a good knowledge of the key players involved in TM speciation is achieved.

Furthermore, development of analytical tools allowing to reach the TM speciation dynamics as defined by Pinheiro et al. [34] under conditions of bulk depletion is needed. In such particular conditions [19], the speciation and bioavailability of TM must be considered at two different time scales: (i) the time scale of the bio-uptake flux, as determined by diffusion of the bioactive free metal, dissociation of the bioinactive complex species, and the internalization rate; and (ii) the time scale of depletion of the bulk medium. Donnan Membrane Technique (DMT) [35] as well as Diffusive Gradient in Thin film (DGT) [36] could provide such type of information provided that the analytical approach allows to work at very low TM concentrations levels as well as in anaerobic conditions. Also a link between TM bioavailability and the microbial community response should be better understood [37].

3 Trace Metal Microbiology and Microbial Ecology

Development and optimisation of AD requires better knowledge of the mechanisms occurring on a microscale, which should in turn be linked to the macroscale system performance and behaviour [38]. Despite this, the relationships between the dynamic behavior of microbial communities and environmental parameters in AD have not been studied in enough detail [1, 39]. A weak component of many AD bioreactor operations is the available information on the structure, dynamics and functions of the microbial community underpinning digestion and biogas production. This is certainly valid for the interactions between TM and the microorganisms.

3.1 Trace Element Requirement in AD

Several studies have reported on the effects of TM deprivation, or supplementation, on the performance of anaerobic bioreactors and on the temporal dynamics of microbial populations. The acidification of methanol-fed bioreactors induced by Co deprivation from the influent was investigated by Fermoso et al. [40] by coupling analysis of bioreactor performance indicators and the microbial ecology, and activity, of the bioreactor sludge. Under Co-limiting conditions in methanol-fed bioreactors, methanogenic activity of bioreactor sludge granules on methanol gradually decreased, leading to methanol accumulation and bioreactor acidification due to acetogenic thermodynamic out-competition of methylothrophic methanogenesis. Methylotrophic, and acetoclastic, methanogenic activity was found to be lost within 10 days of reactor operation, coinciding with the disappearance of a Methanosarcina population. Using fluorescence in situ hybridisations (FISH), along with activity assays, Fermoso et al. [40] concluded that reduced methanogenic activity on methanol, and shifts in population dynamics, could be used as accurate parameters to predict bioreactor acidification under Co-limiting conditions.

Zn-deprivation, and subsequent Zn-supplementation, was investigated in Upflow Anaerobic Sludge Bed (UASB) bioreactors by Fermoso et al. [41]. Significantly reduced methanogenic activity on methanol was observed under Zn-limited conditions, which could not be restored by resuming the continuous supply of the deprived metal. Moreover, FISH analysis indicated that the growth of *Methanosarcina* colonies was irreversibly inhibited by Zn-deprivation. Similarly, Gustavsson et al. [20] found that the microbial community structure in bioreactors treating sulfur-rich feedstocks varied with the availability of Ni and Co. Acetate-utilizing Methanosarcinales were dominant during periods of stable process performance, i.e. with Co and Ni supply, but the abundance of hydrogenotrophic Methanomicrobiales increased significantly, along with volatile fatty acids (VFA) concentrations, under Co or Ni deficiency. The increase was more pronounced at Co limitation. Such studies demonstrate the potential to improve bioreactor performance by managing microbial communities and by supplementation with TM.

However, the impact of TM, and of changing TM concentrations in bioreactors, on biofilm development; microbial community structure; population dynamics; and the metabolism of individual trophic groups and the meta-community (i.e. the microbial community, as a 'meta-organism', incorporating metabolites as well as cells) is still largely unknown. This is a reflection of the inherent, technical challenges facing microbiologists in understanding the reactions and interactions of complex, mixed-species biofilms, as well as the slow shift from viewing dynamic, engineered systems as 'black boxes' without consideration for the active microbial communities present. Nonetheless, it is also related to the obvious focus on reactions and processes converting organic molecules at the expense of understanding the importance and roles of, and requirements for, TM in anaerobic bioreactors.

Verstraete et al. [42] have reviewed the challenge of managing microbial resources in open and complex, mixed-species communities in the context of dynamic populations and chemical micro-environments. They defined the concept of 'Microbial Resource Management' (MRM), which is analogous to Human Resource Management, and the requirement to handle questions such as 'who is there', 'who is doing what with whom' and 'how can one adjust, control and/or steer these mixed cultures and communities'? Just as with the grand challenges of MRM, as outlined by Verstraete et al. [42], such as controlling greenhouse gas emissions from natural environments, or managing the human gut microbiome, similarly provocative questions should be considered as to how the availability of TM impacts the management of microbial communities in anaerobic biotechnologies. The use of TM to manage microbial community structure in the AD context is given support by the study of Feng et al. [37] demonstrating microbial shifts in response to varying concentrations and combinations of TM. For example, they showed that a change of the TM profiles resulted in the occurrence of different strains on *Methanoculleus*. The abundance of these strains varied in relation to concentrations of nickel/molybdenum/boron and/or selenium/tungsten applied in the biogas reactors investigated.

3.2 Assessment of AD Microbial Community Structure and Activity in Relation to Trace Metals

Information on the regulation of TM-responsive gene transcription in microbial species in AD biofilms is obviously also important in understanding and managing the impact of TM, and in optimizing TM application strategies in bioreactors. Kazakov et al. [43], for example, recently described a new family of tungstate-responsive transcriptional regulators in sulfate-reducing bacteria, which are—thermodynamically—important components of the microbial community in many AD bioreactors, where they can compete for substrates and energy with methanogens. However, the impact of responsive regulatory systems on the wider community should also be explored in order to assess the outlook for the performance of whole bioreactors.

Systems biology allows an holistic understanding of the meta-community. Systems biology encompasses information on the DNA sequences; the collectively-transcribed RNA; the translated proteins; and the metabolites resulting from cellular processes. Metagenomics, metatranscriptomics, metaproteomics and metabolomics data can lead from the (i) functional potential of the ecosystem, to

(ii) indications of microbial activity, through (iii) identification of active metabolic pathways and to (iv) identification of the intermediate- and end-products of cellular processes, respectively. However, none—or even pairs—of the 'omics approaches alone can provide comprehensive information on ecosystem function.

Siggins et al. [44] reviewed advances in metaproteomics and the need to combine this with metagenomics and metabolomics. The study of collective expression of all proteins by the individuals in a microbial community, i.e. metaproteomics, can provide insights into microbial functionality. For example, the expression of proteins associated with methanogenic pathways under various states of TM deprivation and supplementation can provide valuable insights to the importance of specific TM for reactions under in situ conditions. However, only collectively can these techniques capture the functional interactions occurring in an ecosystem and track down characteristics that could not be accessed by the study of isolated components.

Successful systems biology strategies will likely be based on the application of polyomics—genomics, transcriptomics, metabolomics and fluxomics—and will benefit also from innovative integrations with techniques and approaches, such as microfluidic cell counting and sorting, and ecological modeling. The objective of systems microbiology in mixed-species communities should be to facilitate prediction of ecosystem characteristics, which, in the context of TM strategies in AD, would support process optimisation with reference to TM concentrations and availability, and the development of new applications. To maximise the impact of systems biology datasets with reference to TM in AD processes, the response and regulation of microbial uptake mechanisms in relation to the speciation of the TM complexes should be focused. Moreover, the impact that the regulatory response of individuals has on the global function of the entire community—and the 'meta-organism' in bioreactors—should be investigated, since inter-dependencies underpin the functioning of diverse AD microbial communities.

In addition, ecosystems biology models have potential for predicting, and hence supporting optimisation and management, of microbial community function in AD systems. Probabilistic models are required, which will incorporate the stochasticity necessary to reflect the environmental conditions in bioreactors that can be used to identify functionally-important groups of microbial individuals in AD systems, and the impact of TM on microorganisms.

Since, ultimately, the goal is to control microbial communities for optimum rates of conversion and bioreactor performance, reliable means are also required of diagnosing problems at the level of the microbial community. This represents a reasonably straightforward and logical requirement once the important, and relevant, pathways and reactions affected by TM availability have been identified. For example, the phylogenetic markers (such as the 16S rRNA gene) or functional genes associated with the populations or processes affected by TM availability can be monitored by quantitative Polymerase Chain Reaction (qPCR) assays. Indeed, gene expression analysis by reverse-transcription quantitative PCR (RT-qPCR), which allows quantification of RNA (transcripts) rather than genes, has been a key enabling technology of the post-genome era. A comprehensive review of the origins and the future potential of qPCR is available from van Guilder et al. [45]. In the context, however, of quantifying DNA or RNA targets from complex microbial communities, and to develop reliable diagnostic tools to support optimized bioreactor operation, including, for example, TM dosing strategies, several additional considerations may apply. Nucleic acids extractions techniques may require optimisation [46] and PCR assays with environmental and bioengineered samples are often particularly challenging [47]. The sensitivity and specificity of qPCR assays, and of the Nucleic Acids diagnostic Tests (NATs) developed thereof, are critical considerations [48]. The cost and reproducibility, as well as the ease-of-use and robustness, of NATs for different samples is also important for optimal use of quantitative, molecular diagnostics tools to monitor the metals-related 'health' of functional groups in anaerobic biotechnologies.

An opportunity to advance this field will be in integrating innovative experimental approaches to measure TM speciation and bioavailability, with community-level microbial ecology and ecophysiology. Furthermore, it is well-known that microbes are able to implement specific strategies to cope with metal deficiency (i.e. micronutrient starvation) or excess (i.e. intrinsic toxicity) by altering the chemical speciation of metals in their surrounding environment. When metals are available in low concentrations or in less bioavailable forms (e.g. under sulfidic condition and extensive metal-sulfide precipitation), microbes are able to excrete strong metal-binding organic compounds to facilitate the uptake of essential metals. These extracellular organic compounds encompass strong metal binding properties and in many cases may result in enhanced solubility of metal-bearing minerals [49]. A well-studied example is microbial Fe acquisition by excretion of Fe-chelating siderophores and further uptake by cognate receptors [50]. Similar processes are believed to be responsible for the increase in Co and Ni solubility and bio-uptake under sulfidic conditions of the natural environments [51, 52]. Some microorganisms have developed metal efflux mechanisms, which pumps excess intracellular metals out of the cell as a strategy against metal toxicity when exposed to high concentrations of metals [53]. Although this topic is well studied in disciplines such as organic chemistry, biogeochemistry and medicine, only a few studies have targeted the effect of microbial stress response under metal deficiency/excess on metal speciation in biogas processes. The various uptake mechanism among the microorganisms active during AD should therefore be targeted as a part of the ecophysiological investigations referred to above aiming at the effects of the TM on the microbial community structure.

4 Biotechnology

Biotechnology links process engineering of anaerobic bioreactors with trace metal speciation and microbial ecology; process engineering decisions will determine the physical and chemical environment for the microbial population, including their access to beneficial trace metals. The operational performance of an anaerobic bioreactor determines the commercial viability of AD installations. A healthy anaerobic bioreactor converts organic material to biogas at the appropriate rate to accommodate the required organic loading and to produce biogas- and hence bioenergy—at a profitable level. It also demonstrates operational stability and resilience and produces digestate that complies with certification limits for farmland application. Figure 2 shows the different aspects of anaerobic digestion that must work cooperatively to yield methane from organic feedstock. The bioavailability of both macro- and micro-elements (including TM) will have an impact on the microbial community and prevailing biochemical pathways in an anaerobic bioreactor, and hence ultimately on the methane produced by that bioreactor.

A poorly performing anaerobic bioreactor might be prone to accumulation of volatile fatty acids leading to acidification and shutdown; or prolonged periods of foaming that make mixing and gas collection difficult and negatively affects the quality of the digestate end-product. For engineers it is important to understand the relationship between TM bioavailability and bioreactor design and operation in order to judge whether TM supplementation would be beneficial to biogas yield. When an anaerobic bioreactor is treating a defined industrial waste or effluent it is fairly straightforward to determine which TM are lacking in the feed; supplementation of the missing elements almost always results in improved biogas yield and operational stability [20, 54], including an interesting example of abatement of

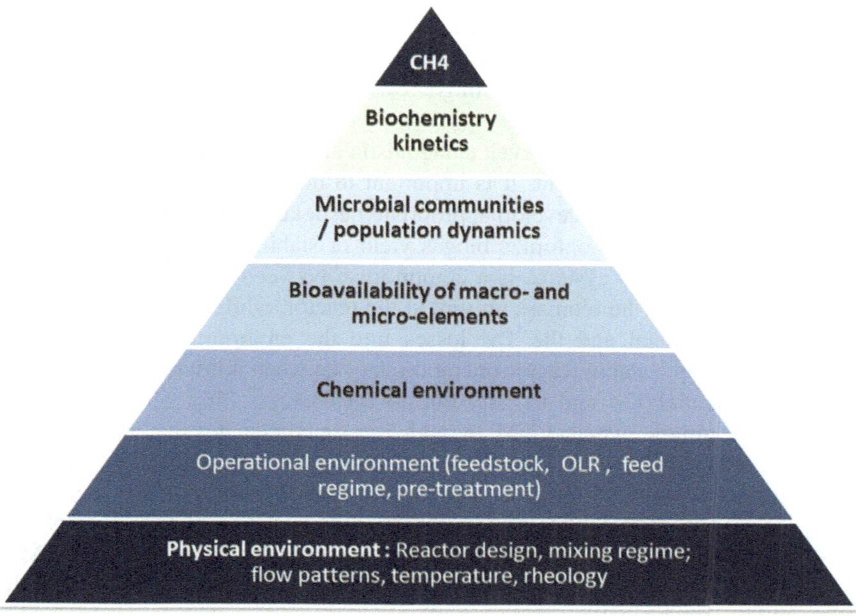

Fig. 2 Depiction of how the physical, operational, chemical and microbiological aspects of anaerobic digestion are involved in the production of methane from organic waste

foaming in a slaughterhouse waste bioreactor supplemented with cobalt (Jörgen Ejlertsson, personal communication).

What is more challenging is to decide whether an anaerobic bioreactor with a full complement of TM in reasonable quantities might in fact be TM deficient; perhaps due to the metals speciated in such a way that they are not bioavailable within the required timescale of digestion. Ishaq et al. [55] showed that sewage sludge bioreactors with no obvious TM deficiencies (but recorded as under-performing with respect to electricity production) responded positively to TM supplementation, in particular cobalt supplementation, with increases in the rate of conversion of acetate to methane being recorded in the range of 9–50 % in Biochemical Methane Potential (BMP) assays.

Required TM supplements are hence determined based on concentration and bioavailability in the feed. The microbial species and hence dominant biochemical pathways in the bioreactor will also determine which TM supplements are most appropriate. For example, Fermoso et al. [40] showed that UASB bioreactors treating methanol were strongly dependent on cobalt supplementation to maintain high biogas yield and bioreactor stability, whereas UASB bioreactors treating an acetate-based waste were less responsive to cobalt supplementation. So a TM supplement that suits one bioreactor will not necessarily be as effective at another bioreactor site. This leads to the question of how the potential benefits of TM supplements can be evaluated at laboratory-scale? BMP batch assays are commonly used to evaluate the effect of TM supplements on the rate of conversion of feedstock to methane. The test is rapid (5–10 days depending on the feedstock) and relatively inexpensive to perform but it evaluates primarily the immediate response of the existing microbial community to TM supplements. The BMP test does not give an indication of how the community might evolve as the TM composition of the bioreactor changes.

TM may have synergistic or even antagonistic effects depending on their relative concentrations in a supplement. It is important to understand which TM are critically limiting to be able to develop appropriate bespoke supplements for individual bioreactors if required to optimize biogas yield or stabilize a bioreactor [15]. TM supplementation in AD systems is a compromise between achieving the maximal biological activity of the biomass present in the reactor, while minimizing the costs of the supplied metal and the TM losses into the environment. The boundary conditions to keep a stable reactor operation vary between nutrient deficiencies due to lack of essential TM and toxicity due to their excess [56]. The TM addition strategy, for example, whether TM are added as metal chelates that are unlikely to precipitate or metal salts that are likely to precipitate, affects the TM losses and hence costs to achieve the optimal TM concentration inside the AD system.

In completely mixed anaerobic reactors, precipitation of TM confers little advantage in terms of TM retention in the bioreactor; hence dosing of TM as soluble (chelated) complexes could be advantageous to improve their availability for rapid microbial uptake. Conversely, anaerobic reactors that decouple the biomass retention time from the hydraulic retention time, such as UASB and SBR (Sequencing Batch Reactor), are likely to benefit from being dosed with TM salts so that the metals will precipitate and accumulate within the biomass sludge fraction, from which they can be assimilated more slowly by the microorganisms as they are required [54, 57]. This is more cost effective than dosing chelated TM that are washed out of the reactor in a matter of hours, even if there is a trade-off in bioavailability.

Depending on the type of substrate to be digested; bioreactor type; and the digestion procedure (mono or co-digestion) employed; TM requirements of anaerobic bioreactors could also theoretically be provided through mixing various feedstocks, such as by co-digesting with sewage sludge, OFMSW or animal excreta, or externally by using chemical additives [58, 59]. Agricultural biogas plants operating with energy crops, such as maize and grass as mono-substrates, can sometimes face suboptimal bioreactor performance without any obvious reason at first glance, but the VFA concentrations range between 3 and 5 kg m^{-3} (recently studied in Germany and Austria by Hinken et al., Pobeheim et al. and Lindorfer et al. [60–62]). Then, the lack or unavailability of micro-nutrients (i.e. trace metals) should be the first reason to be questioned [13].

Anaerobic bioreactors rarely exist in isolation and are usually part of bigger installations with other operations upstream that will have an impact on the bioreactor, for example, sewage sludge anaerobic bioreactors treat the sludge generated through primary and secondary sewage treatment; agricultural anaerobic bioreactors treat manure generated by upstream processes such as milk production; and industrial anaerobic bioreactors are vulnerable to upstream changes in the factory processes that then affect the wastewater composition. Any changes upstream can affect both the composition and bioavailability of TM in the anaerobic bioreactor feedstock and hence have unintended consequences on the methane yield of that bioreactor. A study by Carliell-Marquet et al. [63] showed that introducing iron dosing for chemical phosphorus removal at a sewage treatment works changed the way trace metals were fractionated. Iron shifted to less easily extractable fractions, indicating a potential decrease in bioavailability, whereas copper and zinc moved down the fractionation profiles to be recovered more readily, indicating a potential shift towards being more bioavailable. A key point for AD engineers is that upstream process decisions can impact on TM composition in anaerobic bioreactor feedstock and hence on methane yield. Understanding how process engineering, trace metal chemistry and microbial ecology work together for the benefit of the reactor will enable engineers to plan for and avoid unintended trace metal limitation.

5 Fate of Trace Metals in the Environment

During AD of substrates such as sediments and organic materials, TM species are formed under reducing conditions. Microbial organic matter synthesis effectively reduces redox potential, and produces a range of organic TM complexes and stable inorganic precipitates. After anaerobic digestion, TM are released into the environment as soil or compost via different routes of entry (i.e. suspended /dissolved in effluents or in solid as excess biomass). Once outside the reducing conditions of AD fermenters, waste materials face aerobic conditions that may lead to chemical and physical oxidation and changes in availability of TM. Since legislation has become aware of the potential environmental risks involved (i.e., elevated exposure of (toxic) TM to humans, cattle and ecosystems), the application on land and other forms of re-use is restricted in many EU countries. Nevertheless, the possibilities to include bioavailability in site-specific risk assessment have been included in European guidelines as second-tier methods. This may offer opportunities to derive low-risk conditions and enable safe re-use of materials.

5.1 Risk Assessment of Substrates on Land

Many authors have reported on oxidation mechanisms that occur when anaerobic materials become aerobic [64–67]. The sequence in which oxidation reactions take place have been discussed extensively in terms of thermodynamic pathways, and there is sufficient agreement on the (bio)chemical mechanisms that occur upon reduction and oxidation. Numerical models were developed specifically to address the quantitative effects of redox sequence and TM speciation (e.g., [68, 69]) in terms of resulting pore water concentrations and emission of TM to the environment.

Despite this general agreement on chemical and biochemical mechanisms, the final result of oxidation seldom follows generic rules and often remains uncertain. The final endpoint is the sum of intermediate reactions that occur during oxidation and is largely dictated by the solid phase composition of the substrate and its ability to counteract adverse effects [70–73]. In Table 1, some of these reactions are summarized.

Figure 3 shows an example of the oxidation of anaerobic sediment sludge that was brought on land. This case shows that a drop in pH occurred of almost two units. This is most probably the result of oxidation of relatively large quantities of sulfides, given the increase of SO_4^{2-}. Sulfide-associated TM are released as a consequence. The effect may be superimposed by competition of H^+ ions with TM for organic and inorganic sorption sites.

Xiang et al. [67] used these chemical characteristics to actively release TM from anaerobically digested sludge. The inoculation of indigenous Fe-oxidizing bacteria and the addition of $FeSO_4$ accelerated the solubility of Cr, Cu, Zn, Ni and Pb at a pH that dropped to 2.5. Removal efficiencies were obtained that ranged from 16 % (Pb) to 92 % (Cu). It was reported that the residual TM content in the leached sludge was acceptable for unrestricted use for agriculture.

However, release of H^+ not necessarily has to lead to actual acidification. A substrate or the receiving environment may be capable of buffering the input of protons either by sorption (e.g., by organic components) or reaction with alkaline

Table 1 Major oxidation reactions in digestates and their potential environmental effects

Reaction	Physical or chemical effect
Organic metabolism	Eh ↓; pH ↓; pCO_2 ↑; Possible acidification; Release of sorbed TM; Increase of DOC.
$CH_2O + O_2 \rightarrow CO_2 + H_2O$	
$CO_2 + H_2O \rightarrow HCO_3^- + H^+$	
Sulfide/pyrite oxidation	pH ↓; Release of S-associated TM; Sorption of TM to iron(hydr)oxides; Possible acidification and eutrophication.
$FeS(s) + 9/4O_2 + 1.5H_2O \rightarrow FeOOH(s) + 2H^+ + SO_4^{2-}$	
Nitrification	pH ↓; Possible acidification and eutrophication.
$NH_4^+ + 2O_2 \rightarrow NO_3^- + 2H^+ + H_2O$	

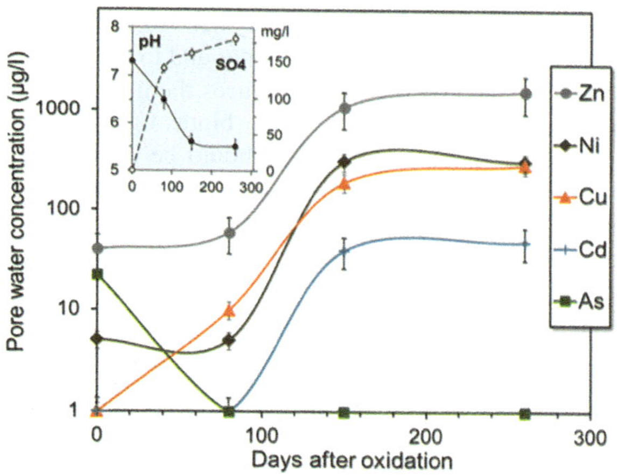

Fig. 3 Examples of pore water concentrations of TM and arsenic after dumping of anaerobic sludge on land (Vink et al., unp.). Oxidation of stable sulfide precipitates releases TM and sulfate. When insufficiently buffered, acidification may give rise to significant release of TM to pore water and possible emission to the surrounding environment

products (e.g., $CaCO_3$). In cases of abundant iron, the formation of reactive iron-oxyhydroxides may drastically increase sorption capacity of the substrate for TM and prevent their release. Using these properties, either intrinsic or by additions, may prove to be a cost-effective and environmentally safe option when compared to active removal processes.

5.2 *Risks Assessment in Aquatic Environments*

For TM, environmental quality standards for sediments and surface waters have been developed to protect the ecosystem from adverse effects. These quality standards are generic, which means that they apply to all surface waters. The importance of explicitly considering bioavailability in the development of water and sediment quality criteria for TM has been recognized for some time [74]. Criteria that incorporate this concept were considered for regulatory implementation for some time [75–77].

Biotic ligand models (BLM) were developed to incorporate local bioavailability in risk assessment procedures. The conceptual framework for the BLM is an adaptation of the gill surface interaction model, originally proposed by Pagenkopf [78, 79] and more recently utilized by many others (e.g., [80–84]), and the free ion activity model of toxicity, extensively reviewed by e.g., [85, 86]. The general framework is illustrated in Fig. 4. The model is based on the hypothesis that toxicity (expressed as No-effect concentration; NOEC) is not simply related to total aqueous TM concentration but that both TM–ligand complexation and TM interaction with competing cations at the site of action of toxicity need to be considered [78, 87, 88]. Mortality occurs when the concentration of TM bound to the biotic ligand exceeds a threshold concentration. The BLM simply replaces the fish gill as the site of action with a more generally characterized site, the biotic ligand. The reason for this replacement is to emphasize that this model should be applicable to other aquatic organisms, like crustaceans, for which the site of action is not readily accessible to direct measurement. It is likely that these principles apply to any organism for which the site of action is directly in contact with the external aqueous environment.

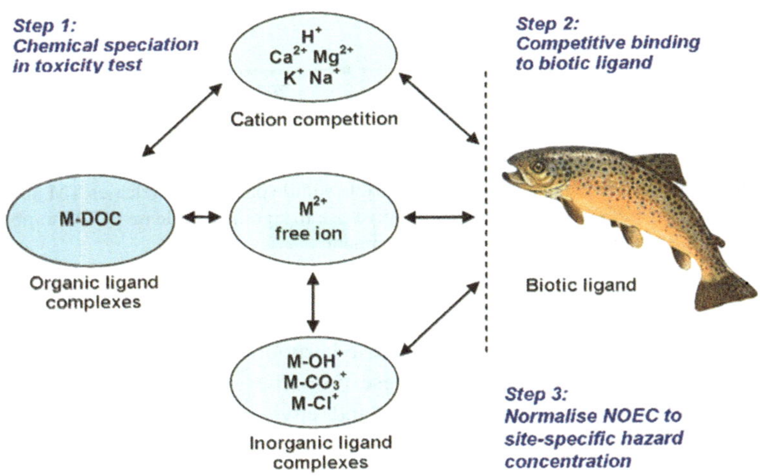

Fig. 4 The biotic ligand model for TM includes calculation of chemical speciation, binding to biota and a normalization procedure to calculate site-specific quality standards (http://www.pnec-pro.com)

Fig. 5 Effect of environmental parameters (Ca, DOC) on toxicity of zinc. The predicted No-effect concentrations are expressed in ug/L. A low PNEC represents a high toxicity; red color indicates zones where the HC5 (hazard concentration for 95 % of species) is approached or exceeded

The role of TM complexation is critical because formation of organic and inorganic metal complexes renders a significant fraction of the total TM non-bioavailable. In fact, this modeling framework defines bioavailability of TM. Dissolved TM exists in solution partially as free TM ion [89, 90]. This species is hypothesized to be the bioavailable species in more simplified versions of the free ion activity model of toxicity. The rest of the TM exists as non-bioavailable TM complexes that result from reactions of the TM with organic and inorganic ligands. Biotic ligand models were developed and validated for Cu, Ni and Zn. Efforts to develop BLMs for other TM (such as Co (e.g., [83]) and Cd (e.g., [91]) are undertaken and tested for uncertainty [92]. For Cd, a bioavailability correction factor based on water hardness has been suggested (e.g., [87]).

Figure 5 visualizes the effects of environmental variability on the actual toxicity of zinc. A large set containing over 80,000 water quality monitoring data were used to calculate the No-effect concentration of zinc under the occurring conditions. As an example, the simultaneous effect of calcium and dissolved organic carbon (DOC) is shown.

Recently, Verschoor et al. [93] performed a geographical and temporal analysis of TM in surface waters, and concluded that "sensitive" and "robust" waters occur, based on their physic-chemical composition. By incorporating chemical speciation of TM in the assessment of ecotoxicological risks for aquatic species, site-specific variations are made visible that remain undetected when checked with generic quality standards. A long history of experiments has provided sufficient evidence for the importance of water chemistry on the adverse effects (toxicity) of heavy metals and other compounds to aquatic organisms. This insight can be used to focus on the protection of vulnerable sites that are identified in this way. Simultaneously, it opens opportunities to explore the possibilities of re-use of digestates in the environment in a responsible manner.

6 Mathematical Modelling

During recent decades, several researchers published a series of mathematical models to simulate the AD process [94–100]. These models are focused on different aspects of AD and have shown good performances in terms of simulation accuracy. The different approaches proposed by the different models, however, make it very difficult to either compare the results of, or to integrate, two or more models.

In 2002, the International Water Association (IWA) Task Group for Mathematical Modelling of Anaerobic Digestion Processes developed a comprehensive mathematical model known as ADM1-Anaerobic Digestion Model no. 1 [101], which was based on the collective knowledge of modelling and simulation of AD systems.

The aim of the ADM1 approach is not to provide an overall model but to supply a unified basis for AD modelling; indeed the first version of ADM1 neglects certain processes involved in AD, such as sulfate reduction, acetate oxidation, homoacetogenesis, solids precipitation and inhibition due to sulfide, nitrate, long-chain fatty acids (LCFAs), and weak acids and bases [102].

Some of the neglected aspects have since been studied and modelled; for instance, two ADM1 extensions were published in 2003 concerning, respectively, sulfate reduction [103] and $CaCO_3$ precipitation [104]. A further extension to remove the ADM1 discrepancies in both carbon and nitrogen balances was later published [105]. Lubken et al. [106] proposed a first modified version of the ADM1, able to take into account the co-digestion of different substrates. More precisely, in the model of Lubken et al. [106], the energy production by co-digesting cattle manure and energy crops was evaluated. Esposito et al. [107] modified the ADM1 to include the possibility to model the disintegration of two different input substrates. Their proposed model considers first-order kinetics for sewage sludge disintegration and surface-based kinetics to model OFMSW disintegration. This model has also since been upgraded [108, 109] to simulate the effect of LCFA production in pH prediction and to include the possibility of separating each product of the disintegration process (i.e. carbohydrates, proteins and lipids) into two fractions, i.e. a readily biodegradable fraction and a slowly biodegradable fraction. Another ADM1 extension published by Barrera et al. [110] in 2015 included sulfate reduction for a very high strength and sulfate rich wastewater. On the model of Barrera et al. [110] not effect of TM was included.

Nevertheless there are several studies that demonstrated the significant effects of TM, and their speciation, in AD. Shakeri Yetka et al. [18] have suggested to combine available bio-uptake theories such as free ion activity and biotic ligand models with the chemical speciation modelling and their further incorporation into mechanistic models describing overall anaerobic digestion process such as ADM1. However, to the best of our knowledge, there are no dynamic mathematical models that could take into account these phenomena. In addition, the ADM1 model does not yet consider the phylogenetic complexity of microbial communities underpinning the AD process, or the dynamic nature of microbial community structure in response to changing environmental conditions. Therefore, a new theoretical model, based on the ADM1 approach, is needed to simulate the effect of TM and their speciation in anaerobic digestion systems.

This mathematical model should include the following characteristics:

1. Ordinary differential equations able to describe the speciation of the TM present in the simulated biological system. In particular these equations should provide the dynamic concentrations of each TM species. Proper description of the TM chemical speciation in a system where diverse organic molecules are present in high concentrations requires identification and inclusion of main metal-binding ligands in particular dissolved organic matter and their stability constants.
2. Differential mass balance equations for substrates, products and bacterial groups involved in the process. These equations include biochemical reactions of substrate conversion and microbial growth and decay kinetics, considering the effects of TM speciation (Table 2). In particular the kinetic equations reported in Table 2 consider the mechanisms of microbial uptake of TM (including their inhibitory and nutritious effects) which is the bridge for connecting TM speciation to microbial growth, decay, and maintenance activities.
3. A module aimed at the determination of the redox potential that is needed to model the TM speciation.
4. Ideally, the incorporation of data on the diversity and abundance, as well as responses to TM concentrations and availability, of the microbial populations in AD bioreactors.
5. Differential parabolic equations capable to model the diffusivity phenomena in the biofilm in case of attached growth reactors. These equations will consider the different diffusivity constant of each TM species.
6. A module to simulate the fate of trace metals in the environment and to assess the risks in soil and aquatic systems.

Table 2 Inhibition forms as used in the ADM1 [101], readapted for TM concentrations. K_I = inhibition parameter; r_j = r ate for process j; S_j = substrate for process j; M_I = inhibitor concentration; X_j = biomass for process j; K_m = Monod maximum specific uptake rate; K_s = half saturation value; Y = yield of biomass on substrate; K_q = first order decay rate; M_{LL} and M_{UL} = lower and upper limits where the group of organisms is 50 % inhibited, respectively

Description	Equation
Uncompetitive inhibition	$\rho_j = \frac{K_m XS}{K_s + S\left(1 + \frac{K_I}{M_I}\right)}$
Competitive inhibition	$\rho_j = \frac{K_m XS}{K_s\left(1 + \frac{M_I}{K_I}\right) + S}$
Reduction in yield	$Y = f(M_I)$
Increased biological decay rate	$K_{dec} = f(M_I)$
Empirical upper and lower inhibition	$I = \frac{1 + a \times b(M_{LL} - M_{UL})}{1 + b(M_I - M_{UL}) + b(M_{LL} - M_I)}$

7 Outlook

A challenging area of AD research remains largely uncharted with respect to understanding the role of TM in enhancing biogas production. Since performance dictates the commercial viability of AD, more attention to managing the TM needs of AD processes will improve the overall prospects for the technology. This major knowledge gap and scientific challenge is a multifaceted problem involving TM chemistry, physical interactions of TM and solids, microbiology, microbial ecology and technology optimization. Moreover, the fate of TMs, and the chemical speciation, transport and ecological impact of TMs in environments—often agricultural lands—receiving discharge waters from AD is largely unknown or at least uncertain. This means that the fate of TM in AD biotechnologies has an inherent importance, which should be explored jointly by multi-disciplinary efforts to provide relevant, and reliable, information and tools for industry and to develop the role and use of TM to enhance biogas production in anaerobic biotechnologies.

Acknowledgments The authors acknowledge funding within the framework of the COST Action 1302 ('European Network on Ecological Roles of Trace Metals in Anaerobic Biotechnologies'). GC is supported by a European Research Council Starting Grant ('3C-BIOTECH; No. 261330).

References

1. Appels L, Lauwers J, Degrve J, Helsen L, Lievens B, Willems K, Van Impe J, Dewil R (2011) Anaerobic digestion in global bio-energy production: Potential and research challenges. Renew Sustain Energy Rev 15(9):4295–4301
2. Lettinga G (2005) The anaerobic treatment approach towards a more sustainable and robust environmental protection, p 1–11
3. Verstraete W, Van de Caveye P, Diamantis V (2009) Maximum use of resources present in domestic "used water". Bioresour Technol 100(23):5537–5545
4. Zeeman G, Kujawa K, de Mes T, Hernandez L, de Graaff M, Abu-Ghunmi L, Mels A, Meulman B, Temmink H, Buisman C, van Lier J, Lettinga G (2008) Anaerobic treatment as a core technology for energy, nutrients and water recovery from source-separated domestic waste(water), p 1207–1212
5. Nichols CE (2004) Overview of anaerobic digestion technologies in Europe. BioCycle, 2004. 45(1):47–48 + 50–53
6. Cameron I (2007) Biogas market comes on strong. Diesel Gas Turbine Worldwide 39(5):10–13
7. Chong S, Sen TK, Kayaalp A, Ang HM (2012) The performance enhancements of upflow anaerobic sludge blanket (UASB) reactors for domestic sludge treatment—A State-of-the-art review. Water Res 46(11):3434–3470
8. Chanakya HN, Malayil S (2012) Anaerobic digestion for bioenergy from agro-residues and other solid wastes—An overview of science, technology and sustainability. J Indian Inst Sci 92(1):111–143
9. Zandvoort MM, van Hullebusch ED, Fermoso FG, Lens P (2006) Trace metals in anaerobic granular sludge reactors: Bioavailability and dosing strategies. Eng Life Sci 6(3):293–301
10. Glass JB, Orphan VJ (2012) Trace metal requirements for microbial enzymes involved in the production and consumption of methane and nitrous oxide. Frontiers in Microbiology, 3

11. Worm P, Fermoso FG, Lens PNL, Plugge CM (2009) Decreased activity of a propionate degrading community in a UASB reactor fed with synthetic medium without molybdenum, tungsten and selenium. Enzyme Microb Technol 45(2):139–145
12. Zandvoort MH, van Hullebusch ED, Gieteling J, Lens PNL (2006) Granular sludge in full-scale anaerobic bioreactors: Trace element content and deficiencies. Enzyme Microb Technol 39(2):337–346
13. Demirel B, Scherer P (2011) Trace element requirements of agricultural biogas digesters during biological conversion of renewable biomass to methane. Biomass Bioenergy 35 (3):992–998
14. Facchin V, Cavinato C, Pavan P, Bolzonella D (2013) Batch and continuous mesophilic anaerobic digestion of food waste: Effect of trace elements supplementation. Chem Eng Trans 32:457–462
15. Banks CJ, Zhang Y, Jiang Y, Heaven S (2012) Trace element requirements for stable food waste digestion at elevated ammonia concentrations. Bioresour Technol 104:127–135
16. van Hullebusch ED, Zandvoort MH, Lens PNL (2003) Metal immobilisation by biofilms: Mechanisms and analytical tools. Rev Environ Sci Biotechnol 2(1):9–33
17. Möller K, Müller T (2012) Effects of anaerobic digestion on digestate nutrient availability and crop growth: A review. Eng Life Sci 12(3):242–257
18. Yekta SS, Svensson BH, Björn A, Skyllberg U (2014) Thermodynamic modeling of iron and trace metal solubility and specia-tion under sulfidic and ferruginous conditions in full scale continuous stirred tank biogas reactors. Appl Geochem (In press)
19. Jansen S (2004) Speciation and bioavailability of cobalt and nickel in anaerobic wastewater treatment. In: Leerstoelgroep Fysische Chemie en Kolloidkunde. 2004, Wageningen University
20. Gustavsson J, Yekta SS, Sundberg C, Karlsson A, Ejlertsson J, Skyllberg U, Svensson BH (2013) Bioavailability of cobalt and nickel during anaerobic digestion of sulfur-rich stillage for biogas formation. Appl Energy 112:473–477
21. Zandvoort MH, van Hullebusch ED, Gieteling J, Lettinga G, Lens PNL (2005) Effect of sulfur source on the performance and metal retention of methanol-fed UASB reactors. Biotechnol Prog 21(3):839–850
22. Yekta SS, Lindmark A, Skyllberg U, Danielsson T, Svensson BH (2014) Importance of reduced sulfur for the equilibrium chemistry and kinetics of Fe(II), Co(II) and Ni(II) supplemented to semi-continuous stirred tank biogas reactors fed with stillage. J Hazard Mater 269:83–88
23. Kenney JPL, Fein JB (2011) Cell wall reactivity of acidophilic and alkaliphilic bacteria determined by potentiometric titrations and cd adsorption experiments. Environ Sci Technol 45(10):4446–4452
24. d'Abzac P, Bordas F, van Hullebusch ED, Lens PNL, Guibaud G (2010) Effects of extraction procedures on metal binding properties of extracellular polymeric substances (EPS) from anaerobic granular sludges. Colloids Surf B 80(2):161–168
25. Aquino SF, Stuckey DC (2007) Bioavailability and toxicity of metal nutrients during anaerobic digestion. J Environ Eng 133(1):28–35
26. Gonzalez-Gil G, Jansen S, Zandvoort M, van Leeuwen HP (2003) Effect of yeast extract on speciation and bioavailability of nickel and cobalt in anaerobic bioreactors. Biotechnol Bioeng 82(2):134–142
27. Bartacek J, Fermoso FG, Baldo-Urrutia AM, van Hullebusch ED, Lens PNL (2008) Cobalt toxicity in anaerobic granular sludge: Influence of chemical speciation. J Ind Microbiol Biotechnol 35(11):1465–1474
28. Fermoso FG, Bartacek J, Chung LC, Lens P (2008) Supplementation of cobalt to UASB reactors by pulse dosing: CoCl2 versus CoEDTA2- pulses. Biochem Eng J 42(2):111–119
29. d'Abzac P, Bordas F, Joussein E, van Hullebusch ED, Lens PNL, Guibaud G (2013) Metal binding properties of extracellular polymeric substances extracted from anaerobic granular sludges. Environ Sci Pollut Res 20(7):4509–4519
30. Yekta SS, Gustavsson J, Svensson BH, Skyllberg U (2012) Sulfur K-edge XANES and acid volatile sulfide analyses of changes in chemical speciation of S and Fe during sequential extraction of trace metals in anoxic sludge from biogas reactors. Talanta 89:470–477

31. Li X, Dai X, Takahashi J, Li N, Jin J, Dai L, Dong B (2014) New insight into chemical changes of dissolved organic matter during anaerobic digestion of dewatered sewage sludge using EEM-PARAFAC and two-dimensional FTIR correlation spectroscopy. Bioresour Technol 159:412–420
32. Van der Veen A, Fermoso FG, Lens P (2007) Bonding form analysis of metals and sulfur fractionation in methanol-grown anaerobic granular sludge engineering in life. Science 7 (5):480–489
33. van Hullebusch ED, Rossano S, Farges F, Lenz M, Labanowski J, Lagarde P, Flank A-M, Lens PNL (2009) Sulfur K-edge XANES spectroscopy as a tool for understanding sulfur chemical state in anaerobic granular biofilms. In: 14th international conference on X-ray absorption fine structure (XAFS14), Camerino, Italy
34. Pinheiro JP, Galceran J, Van Leeuwen HP (2004) Metal speciation dynamics and bioavailability: bulk depletion effects. Environ Sci Technol 38(8):2397–2405
35. Temminghoff EJM, Plette ACC, Van Eck R, Van Riemsdijk WH (2000) Determination of the chemical speciation of trace metals in aqueous systems by the Wageningen Donnan membrane technique. Anal Chim Acta 417(2):149–157
36. Davison W, Zhang H (1994) In situ speciation measurements of trace components in natural waters using thin-film gels. Nature 367(6463):546
37. Feng XM, Karlsson A, Svensson BH, Bertilsson S (2010) Impact of trace element addition on biogas production from food industrial waste—Linking process to microbial communities. FEMS Microbiol Ecol 74(1):226–240
38. Rittmann BE, Krajmalnik-Brown R, Halden RU (2008) Pre-genomic, genomic and post-genomic study of microbial communities involved in bioenergy. Nat Rev Microbiol 6 (8):604–612
39. Tomei MC, Braguglia CM, Cento G, Mininni G (2009) Modeling of Anaerobic digestion of sludge. Crit Rev Environ Sci Technol 39(12):1003–1051
40. Fermoso FG, Collins G, Bartacek J, O'Flaherty V, Lens P (2008) Acidification of methanol-fed anaerobic granular sludge bioreactors by cobalt deprivation: induction and microbial community dynamics. Biotechnol Bioeng 99(1):49–58
41. Fermoso FG, Collins G, Bartacek J, Lens PNL (2008) Zinc deprivation of methanol fed anaerobic granular sludge bioreactors. J Ind Microbiol Biotechnol 35(6):543–557
42. Verstraete W, Wittebolle L, Heylen K, Vanparys B, de Vos P, van de Wiele T, Boon N (2007) Microbial resource management: the road to go for environmental biotechnology. Eng Life Sci 7(2):117–126
43. Kazakov AE, Rajeev L, Luning EG, Zane GM, Siddartha K, Rodionov DA, Dubchak I, Arkin AP, Wall JD, Mukhopadhyay A, Novichkov PS (2013) New family of tungstate-responsive transcriptional regulators in sulfate-reducing bacteria. J Bacteriol 195 (19):4466–4475
44. Siggins A, Gunnigle E, Abram F (2012) Exploring mixed microbial community functioning: recent advances in metaproteomics. FEMS Microbiol Ecol 80(2):265–280
45. VanGuilder HD, Vrana KE, Freeman WM (2008) Twenty-five years of quantitative PCR for gene expression analysis. Biotechniques 44(5):619–626
46. Carrigg C, Rice O, Kavanagh S, Collins G, O'Flaherty V (2007) DNA extraction method affects microbial community profiles from soils and sediment. Appl Microbiol Biotechnol 77 (4):955–964
47. Smith CJ, Osborn AM (2009) Advantages and limitations of quantitative PCR (Q-PCR)-based approaches in microbial ecology. FEMS Microbiol Ecol 67(1):6–20
48. Myint MS, Johnson YJ, Tablante NL, Heckert RA (2006) The effect of pre-enrichment protocol on the sensitivity and specificity of PCR for detection of naturally contaminated Salmonella in raw poultry compared to conventional culture. Food Microbiol 23(6):599–604
49. Banfield JF, Barker WW, Welch SA, Taunton A (1999) Biological impact on mineral dissolution: Application of the lichen model to understanding mineral weathering in the rhizosphere. Proc Natl Acad Sci USA 96(7):3404–3411

50. Krewulak KD, Vogel HJ (2008) Structural biology of bacterial iron uptake. Biochim Biophys Acta Biomembr 1778(9):1781–1804
51. Saito MA, Moffett JW, Chisholm SW, Waterbury JB (2002) Cobalt limitation and uptake in Prochlorococcus. J Limnol Oceanogr 47(6):1629–1636
52. Waldron KJ, Robinson NJ (2009) How do bacterial cells ensure that metalloproteins get the correct metal? Nat Rev Microbiol 7(1):25–35
53. Rodrigue A, Effantin G, Mandrand-Berthelot MA (2005) Identification of rcnA (yohM), a nickel and cobalt resistance gene in Escherichia coli. J Bacteriol 187(8):2912–2916
54. Fermoso FG, Bartacek J, Manzano R, van Leeuwen HP, Lens PN (2010) Dosing of anaerobic granular sludge bioreactors with cobalt: impact of cobalt retention on methanogenic activity. Bioresour Technol 101(24):9429–9437
55. Ishaq F, Bridgeman J, Carliell-Marquet CM (2013) Site energy performance as an indicator for trace element deficiency in full-scale digesters. In: 13th world congress on anaerobic digestion (IWA specialist conference). Santiago de Compostela, Spain
56. Fermoso FG, Bartacek J, Jansen S, Lens PN (2009) Metal supplementation to UASB bioreactors: from cell-metal interactions to full-scale application. Sci Total Environ 407 (12):3652–3667
57. Jansen S, Gonzalez-Gil G, van Leeuwen HP (2007) The impact of Co and Ni speciation on methanogenesis in sulfidic media - Biouptake versus metal dissolution. Enzyme Microb Technol 40(4):823–830
58. Nges IA, Björnsson L (2012) High methane yields and stable operation during anaerobic digestion of nutrient-supplemented energy crop mixtures. Biomass Bioenergy 47:62–70
59. Nges IA, Björn A, Björnsson L (2012) Stable operation during pilot-scale anaerobic digestion of nutrient-supplemented maize/sugar beet silage. Bioresour Technol 118:445–454
60. Pobeheim H, Munk B, Lindorfer H, Guebitz GM (2011) Impact of nickel and cobalt on biogas production and process stability during semi-continuous anaerobic fermentation of a model substrate for maize silage. Water Res 45(2):781–787
61. Lindorfer H, Ramhold D, Frauz B (2012) Nutrient and trace element supply in anaerobic digestion plants and effect of trace element application. Water Sci Technol 66(9):1923–1929
62. Hinken L, Urban I, Haun E, Weichgrebe D, Rosenwinkel KH (2008) The valuation of malnutrition in the mono-digestion of maize silage by anaerobic batch tests. Water Sci Technol 58:1453–1459
63. Carliell-Marquet C, Smith J, Oikonomidis I, Wheatley A (2010) Inorganic profiles of chemical phosphorus removal sludge. Proc Inst Civil Eng Water Manag 163(2):65–77
64. Holmes J (1999) Fate of incorporated metals during mackinawite oxidation in sea water. Appl Geochem 14(3):277–281
65. Mayer TD, Jarrell WM (2000) Phosphorus sorption during iron(II) oxidation in the presence of dissolved silica. Water Res 34(16):3949–3956
66. Simpson SL, Apte SG, Batley GE (2000) Effect of short-term resuspension events on the oxidation of cadmium, lead, and zinc sulfide phases in anoxic estuarine sediments. Environ Sci Technol 34(21):4533–4537
67. Xiang L, Chan LC, Wong JWC (2000) Removal of heavy metals from anaerobically digested sewage sludge by isolated indigenous iron-oxidizing bacteria. Chemosphere 41(1–2):283–287
68. Vink JPM, Meeussen JCL (2007) BIOCHEM-ORCHESTRA: A tool for evaluating chemical speciation and ecotoxicological impacts of heavy metals on river flood plain systems. Environ Pollut 148(3):833–841
69. Christensen JB, Christensen TH (1999) Complexation of Cd, Ni, and Zn by DOC in polluted groundwater: a comparison of approaches using resin exchange, aquifer material sorption, and computer speciation models (WHAM and MINTEQA2). Environ Sci Technol 33 (21):3857–3863
70. Di Toro DM, Allen HE, Bergman HL, Meyer JS, Paquin PR, Santore RC (2001) Biotic ligand model of the acute toxicity of metals. 1. Technical basis. Environ Toxicol Chem 20 (10):2383–2396

71. Lamers LPM, Tomassen HBM, Roelofs JGM (1998) Sulfate-induced eutrophication and phytotoxicity in freshwater wetlands. Environ Sci Technol 32(2):199–205
72. Siegrist H, Brunner I, Koch G, Phan LC, Van Le C (1999) Reduction of biomass decay rate under anoxic and anaerobic conditions. Water Sci Technol 39:129–137
73. Vink JPM, Harmsen J, Rijnaarts H (2010) Delayed immobilization of heavy metals in soils and sediments under reducing and anaerobic conditions; consequences for flooding and storage. J Soils Sediments 10(8):1633–1645
74. Ankley GT, Di Toro DM, Hansen DJ, Berry WJ (1996) Technical basis and proposal for deriving sediment quality criteria for metals. Environ Toxicol Chem 15(12):2056–2066
75. Bergman HL, Dorward-King EJ (1997) Reassessment of metals criteria for aquatic life protection: priorities for research and implementation. In: Proceedings of the pellston workshop on reassessment of metals criteria for aquatic life protection, pp 10–14 February 1996, Pensacola, Florida, SETAC Press
76. Renner R (1997) Rethinking water quality standards for metals toxicity. Environ Sci Technol 31(10):465A–468A
77. EU (2008) Directive 2008/105/EC of the European Parliament and of the Council of 16 December 2008 on environmental quality standards in the field of water policy, amending and subsequently repealing Council Directives 82/176/EEC, 83/513/EEC, 84/156/EEC, 84/491/EEC, 86/280/EEC and amending Directive 2000/60/EC of the European Parliament and of the Council, p 14
78. Pagenkopf GK (1983) Gill surface interaction model for trace-metal toxicity to fishes: Role of complexation, pM, and water hardness. Environ Sci Technol 17(6):342–347
79. Pagenkopf GK, Russo RC, Thurston RV (1974) Effect of complexation on toxicity of copper to fishes. J Fish Res Board Canada 31(4):462–465
80. Hollis L, Burnison K, Playle RC (1996) Does the age of metal-dissolved organic carbon complexes influence binding of metals to fish gills? Aquat Toxicol 35(3–4):253–264
81. Janes N, Playle RC (1995) Modeling silver binding to gills of rainbow trout (Oncorhynchus mykiss). Environ Toxicol Chem 14(11):1847–1858
82. Playle RC (1998) Modelling metal interactions at fish gills. Sci Total Environ 219(2–3):147–163
83. Richards JG, Playle RC (1998) Cobalt binding to gills of rainbow trout (Oncorhynchus mykiss): An equilibrium model. Comp Biochem Physiol C Pharmacol Toxicol Endocrinol 119(2):185–197
84. Wood CM, Playle RC, Hogstrand C (1999) Physiology and modeling of mechanisms of silver uptake and toxicity in fish. Environ Toxicol Chem 18(1):71–83
85. Morel F (1983) Principles of aquatic chemistry. Wiley, New York
86. Campbell PG (1995) In Interactions between trace metals and aquatic organ-isms: A critique of the free-ion activity model, in Metal Speciation and Bioavailability in AquaticSystems. In: Tessier A, Turner D (eds), John Wiley: New York, NY, USA. p. 45–102
87. Meyer JS (1999) A mechanistic explanation for the In(LC50) vs In(hardness) adjustment equation for metals. Environ Sci Technol 33(6):908–912
88. Paquin PR, di Toro DM, Santore RC, Trivedi D, Wu KB (1999) A Biotic Ligand Model of the Acute Toxicity of Metals: III. Application to Fish and Daphnia Exposure to Silver. In: Integrated Approach to Assessing the Bioavailability and Toxicity of Metals in Surface Waters and Sediments. U.E. EPA-822-E-99-001, Editor. 1999: Washington, DC, p 3-59–3-102
89. Vink JPM (2009) The origin of speciation: Trace metal kinetics over natural water/sediment interfaces and the consequences for bioaccumulation. Environ Pollut 157(2):519–527
90. Vink JPM (2002) Measurement of heavy metal speciation over redox gradients in natural water-sediment interfaces and implications for uptake by benthic organisms. Environ Sci Technol 36(23):5130–5138
91. Niyogi S, Kent R, Wood CM (2008) Effects of water chemistry variables on gill binding and acute toxicity of cadmium in rainbow trout (Oncorhynchus mykiss): A biotic ligand model (BLM) approach. Comp Biochem Physiol C Toxicol Pharmacol 148(4):305–314

92. Vijver MG, De Koning A, Peijnenburg WJGM (2008) Uncertainty of water type-specific hazardous copper concentrations derived with biotic ligand models. Environ Toxicol Chem 27(11):2311–2319
93. Verschoor AJ, Vink JPM, De Snoo GR, Vijver MG (2011) Spatial and temporal variation of watertype-specific no-effect concentrations and risks of Cu, Ni, and Zn. Environ Sci Technol 45(14):6049–6056
94. Hill D, Barth C (1977) A dynamic model for simulation of animal waste digestion. J (Water Pollut Control Fed) 49:2129–2143
95. Kalyuzhnyi S, Davlyatshina M (1997) Batch anaerobic digestion of glucose and its mathematical modeling. I. Kinetic investigations. Bioresour Technol 59(1):73–80
96. Mosey F (1983) Mathematical modelling of the anaerobic digestion process: regulatory mechanisms for the formation of short-chain volatile acids from glucose. Water Sci Technol 15(8–9):209–232
97. Costello D, Greenfield P, Lee PL (1991) Dynamic modelling of a single-stage high-rate anaerobic reactor—I. Model derivation. Water Res 25(7):847–858
98. Batstone D, Keller J, Newell R, Newland M (2000) Modelling anaerobic degradation of complex wastewater. I: model development. Bioresour Technol 75(1):67–74
99. Angelidaki I, Ellegaard L, Ahring BK (1993) A mathematical model for dynamic simulation of anaerobic digestion of complex substrates: focusing on ammonia inhibition. Biotechnol Bioeng 42(2):159–166
100. Vavilin V, Vasiliev V, Ponomarev A, Rytow S (1994) Simulation model 'methane'as a tool for effective biogas production during anaerobic conversion of complex organic matter. Bioresour Technol 48(1):1–8
101. Batstone DJ, Keller J, Angelidaki I, Kalyuzhnyi SV, Pavlostathis SG, Rozzi A, Sanders WT, Siegrist H, Vavilin VA (2002) The IWA Anaerobic Digestion Model No 1 (ADM1). Water Sci Technol 45(10):65–73
102. Fuentes M, Scenna NJ, Aguirre PA, Mussati MC (2008) Application of two anaerobic digestion models to biofilm systems. Biochem Eng J 38(2):259–269
103. Fedorovich V, Lens P, Kalyuzhnyi S (2003) Extension of anaerobic digestion model no. 1 with processes of sulfate reduction. Appl Biochem Biotechnol Part A Enzyme Eng Biotechnol 109(1–3):33–45
104. Batstone DJ, Keller J (2003) Industrial applications of the IWA anaerobic digestion model No. 1 (ADM1). Water Sci Technol 47:199–206
105. Blumensaat F, Keller J (2005) Modelling of two-stage anaerobic digestion using the IWA Anaerobic Digestion Model No. 1 (ADM1). Water Res 39(1):171–183
106. Lübken M, Wichern M, Schlattmann M, Gronauer A, Horn H (2007) Modelling the energy balance of an anaerobic digester fed with cattle manure and renewable energy crops. Water Res 41(18):4085–4096
107. Esposito G, Frunzo L, Panico A, d'Antonio G (2008) Mathematical modelling of disintegration-limited co-digestion of OFMSW and sewage sludge. Water Sci Technol 58 (7):1513–1519
108. García-Gen S, Sousbie P, Rangaraj G, Lema JM, Rodríguez J, Steyer JP, Torrijos M (2015) Kinetic modelling of anaerobic hydrolysis of solid wastes, including disintegration processes. Waste Manag 35:96–104
109. Esposito G, Frunzo L, Panico A, Pirozzi F (2011) Modelling the effect of the OLR and OFMSW particle size on the performances of an anaerobic co-digestion reactor. Process Biochem 46(2):557–565
110. Barrera EL, Spanjers H, Solon K, Amerlinck Y, Nopens I, Dewulf J (2015) Modeling the anaerobic digestion of cane-molasses vinasse: extension of the Anaerobic Digestion Model No. 1 (ADM1) with sulfate reduction for a very high strength and sulfate rich wastewater. Water Res 71:42–54

Index

G.M. Guebitz et al. (eds.), *Biogas Science and Technology*,
Advances in Biochemical Engineering/Biotechnology 151,
DOI 10.1007/978-3-319-21993-6

The manufacturer's authorised representative in the EU is Springer Nature Customer Service Centre GmbH, Europaplatz 3, 69115 Heidelberg, Germany. If you have any concerns regarding our products, please contact ProductSafety@springernature.com

Printed and bound by CPI Group (UK) Ltd, Croydon, CR0 4YY
25/09/2026
02222855-0003